A DICTIONARY OF GENETIC ENGINEERING

STEPHEN G. OLIVER

Department of Biochemistry and Applied Molecular Biology
University of Manchester Institute of Science and Technology

JOHN M. WARD

Department of Biochemistry, University College, London

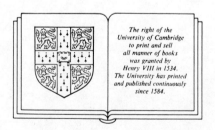

The right of the
University of Cambridge
to print and sell
all manner of books
was granted by
Henry VIII in 1534.
The University has printed
and published continuously
since 1584.

CAMBRIDGE UNIVERSITY PRESS

Cambridge

London New York New Rochelle

Melbourne Sydney

Published by the Press Syndicate of the University of Cambridge
The Pitt Building, Trumpington Street, Cambridge CB2 1RP
32 East 57th Street, New York, NY 10022, USA
10 Stamford Road, Oakleigh, Melbourne 3166, Australia

First published 1985

Printed in Great Britain by the University Press, Cambridge

Library of Congress catalogue card number: 84–19973

British Library Cataloguing in Publication Data
Oliver, S. G.
A dictionary of genetic engineering.
1. Genetic engineering
I. Title II. Ward, John M.
575.1   QH442

ISBN 0 521 26080 9

Contents

Figures

How to use this dictionary

The definitions in this dictionary are arranged in strict alphabetical order.
If you wish to look up a term which includes Greek letters or Roman or
Arabic numerals, you should first transliterate it into Latin script.
For example:

φX174   will be found under p-h-i-x-o-n-e-,
Qβ   will be found under q-b-e-t-a,
λ   will be found under l-a-m-b-d-a,
5′   will be found under f-i-v-e-p-r-i-m-e,
RP4   will be found under r-p-f-o-u-r,
cI   will be found under c-o-n-e.

A

actinomycete A member of the bacterial class, Actinomycetales. These Gram-positive, spore-forming, mycelial bacteria are abundant in soils and composts. Many actinomycetes produce volatile fatty acids which give earth its characteristic odour. In their natural environment they are responsible for the breakdown and recycling of substances such as cellulose, chitin and keratin. Actinomycetes, especially *Streptomyces* species, produce most of the world's antibiotics and many different species are cultivated on a large scale for the commercial production of clinically useful antibiotics. In genetic engineering certain streptomycetes have been used to develop a host-vector system for cloning.

activator (i) In molecular biology, a protein which binds to a site on DNA upstream of a gene and activates transcription from that gene. (ii) In enzymology, a small molecule which binds to an enzyme and increases its catalytic activity.

agarose gel An inert matrix used in the electrophoretic separation of nucleic acid molecules on the basis of their size or conformation. Gels may be cast as tubes or slabs although the latter now predominate. Molecules are visualised in the gel by the ultraviolet fluorescence of ethidium bromide which is either included in the running buffer or used to stain the gel after electrophoresis. (*See* comb, LGT agarose, power pack, Tris-acetate buffer, Tris-borate buffer)

Agrobacterium rhizogenes A species of Gram-negative, rod-shaped soil bacteria closely related to *Agrobacterium tumefaciens*. *A. rhizogenes* often harbours large plasmids, called Ri plasmids, which are closely related to Ti plasmids. The combination of *A. rhizogenes* and an Ri plasmid can cause a tumourous growth known as hairy root disease in certain types of plants.

Agrobacterium tumefaciens A species of soil bacterium which, when it contains a Ti plasmid, can infect the stems of many plants and form crown gall tumours.

agropine A rare amino acid derivative which is produced by a certain type of crown gall tumour. The genes responsible for the synthesis of agropine are part of the T-DNA from a Ti plasmid.

alkaline hydrolysis The use of a high pH to degrade or hydrolyse a bond. DNA is not hydrolysed at high pH while RNA will be degraded to

1

mononucleotides. RNA has a 2′ hydroxyl group which, at high pH, will attack the 3′ phosphodiester bond. DNA has no hydroxyl at the 2′ position and is thus stable to alkaline hydrolysis.

alkaline phosphatase An enzyme which removes the 5′ terminal phosphate groups from linear DNA molecules. It is used to prevent the religation of plasmid vector molecules following cleavage with a restriction endonuclease. This increases the chance that intact circular molecules generated by the ligase reaction are recombinant.

α-peptide A short (185 amino acids) amino-terminal fragment of the enzyme β-galactosidase. The α-peptide can combine with, and restore the activity of, a non-functional β-galactosidase enzyme which has a defective N-terminus. M13mp phage cloning vectors make use of this complementation as they carry the gene for the α-peptide.

amber A mutation which creates the stop codon UAG in the coding region of a gene. This leads to the synthesis of a truncated protein. These mutations can be suppressed by certain mutant tRNA species which permit the incorporation of an amino acid in response to the UAG stop codon thus allowing the protein to be completed. Amber mutations are deliberately incorporated into certain λ phage cloning vectors so that they can only be propagated in hosts which will suppress the amber mutation. This is a type of biological containment. In genetic notation, amber is abbreviated to *am*, thus an amber mutation in gene *S* is written as *Sam*.

Ampicillin resistant Ap^r; Ampicillin sensitive Ap^s Resistance or sensitivity to the lethal effects of the antibiotic ampicillin. Ampicillin is a β-lactam antibiotic and resistance is (often) mediated by a class of enzymes called β-lactamases which are secreted either into the periplasmic space of Gram-negative bacteria or into the medium in Gram-positive bacteria. The cloning vector pBR322 contains an ampicillin resistance gene.

amplification An increase in the copy number of a gene or plasmid. (*See* chloramphenicol amplification)

angle rotor, fixed-angle rotor A centrifuge rotor in which the wells holding the tubes are drilled at an angle to both the axis of rotation and the lines of centrifugal force. Angle rotors were originally used simply to pellet material in the technique of differential centrifugation. They are now used routinely for density gradient centrifugation. The gradients formed in angle rotors are not linear, but permit a great deal of resolution over a narrow density range.

anneal A verb meaning to hybridise nucleic acid molecules.

antibiotic A substance, produced by one organism, which inhibits or kills another organism. Most antibiotics are active against bacteria and are produced by fungi or streptomycetes. (*See* ampicillin resistance, tetracycline resistance)

antibiotic resistance Resistance to the lethal effects of an antibiotic. There are five main mechanisms of antibiotic resistance: (i) inactivation of the antibiotic; (ii) a reduction in cellular uptake or an increase in cellular excretion of the antibiotic; (iii) production of an altered target protein which no longer binds the antibiotic; (iv) overproduction of the target protein such that the antibiotic is 'titrated out'; (v) elaboration of some alternative enzyme or pathway which is not susceptible to the antibiotic.

anticodon The three nucleotides in a tRNA molecule which base pair with the complementary nucleotides forming the codon within mRNA. It is this codon–anticodon interaction, taking place at the ribosome, which ensures that the correct amino acid is inserted into the growing polypeptide chain.

anti-terminator A type of protein which enables RNA polymerase to ignore certain transcriptional stop or termination signals and read through them to produce longer mRNA transcripts.

Ap^r, Ap^s Ampicillin resistant, ampicillin sensitive.

Arabidopsis thaliana Thale Cress. A small, fast-growing dicotyledonous plant, a member of the Cruciferae. It is a favoured experimental organism among plant molecular biologists.

ARS Autonomously replicating sequence (or segment). A term, commonly used in yeast molecular biology, for a DNA sequence which will support independent replication of a plasmid in a host yeast cell. Some *ARS* sequences may be cloned from either the yeast itself or from some other organism. They are thought to represent origins of DNA replication, although it is not clear that they are actually used as such in their parent genomes. Recombinant plasmids which rely on an *ARS* sequence for their replication are intrinsically unstable in yeast (see YRp). The term may, in principle, be applied to sequences which promote plasmid replication in any organism.

A's and T's method A method whereby random DNA fragments can be cloned into a vector molecule. Random DNA fragments generated by mechanically shearing or sonicating genomic DNA are treated with λ-exonuclease in order to generate 3′ single-strand tails. These tails are

3

then extended by the addition of deoxyadenosine residues using the enzyme terminal transferase (calf thymus terminal deoxynucleotidyl transferase). The vector molecule is cut at a unique restriction site to generate 3′ tails again. These tails are extended with deoxythymidine and terminal transferase. The foreign and vector DNA now bear complementary 3′ tails and can be annealed together. If the tails are long enough the molecule will be sufficiently stable not to require treatment with ligase before being used to transform a host organism. While certain advantages attach to the cloning of truly random fragments, the technique has the drawback that it is difficult to retrieve the insert from the recombinant molecule. However, the lower melting temperature of the oligo dA·dT junction sequences may be exploited by partially denaturing and then cutting with a single-strand-specific endonuclease such as S1.

Fig. 1. A's and T's method.

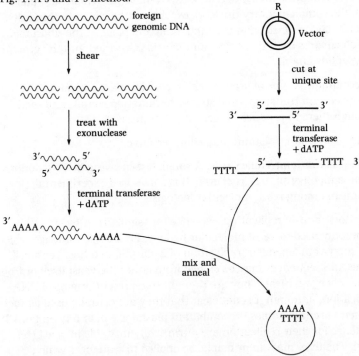

Aspergillus A filamentous fungus of both industrial and genetic importance. Two imperfect (non-sexual) species, *A. niger* and *A. oryzae*, are used in the production of citric acid, industrial enzymes and fermented foods. The perfect (sexual) species *A. nidulans* has been an important research tool in both biochemical and mitochondrial genetics. Its life cycle is given below.

Fig. 2. *Aspergillus nidulans* life cycle.

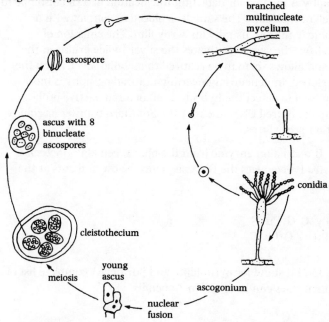

attenuator A sequence which is found upstream of a bacterial operon which encodes enzymes involved in amino acid biosynthesis. The attenuator regulates the expression of such operons by determining whether the mRNA molecules containing their transcripts will be completed or not. The attenuator sequence contains a short open reading-frame which includes several codons specifying the amino acid which the gene products of the regulated operon synthesise. If the concentration of this amino acid in the cell is limiting, then a ribosome will pause in its translation of the attenuator sequence in the nascent message. The continued presence of the ribosome in the attenuator region favours the formation of one of two possible secondary structures in the attenuator transcript. This secondary structure permits the RNA polymerase to extend the mRNA chain to include the operon transcript. If, on the other hand, there is a plentiful supply of the amino acid in the cell then the alternative secondary structure is formed within the attenuator mRNA. This structure is recognised as a terminator by the RNA polymerase and transcription is halted (attenuated) before the operon is transcribed. In this way, mRNA for the amino acid biosynthetic enzymes is only produced when there is a high demand for amino acid biosynthesis.

autoradiography A method for detecting the location of a radioisotope in a tissue, cell or molecule. The sample is placed in contact with a photographic emulsion, usually an X-ray film. The emission of β-particles from the sample activates the silver halide grains in the emulsion and allows them to be reduced to metallic silver when the film is developed. In genetic engineering, autoradiography is most commonly used to detect the hybridisation of a radioactive probe molecule to denatured DNA in either the Southern transfer or colony hybridisation procedures.

Ava I A type II restriction enzyme from the blue-green alga *Anabaena variabilis*. *Ava* I recognises the DNA sequence below and cuts at the sites shown by the arrows:

     ↓
5′ C Py C G Pu G 3′
3′ G Pu G C Py C 5′
               ↑

where Py and Pu stand for pyrimidine and purine. (A complete list of restriction enzymes can be found in Appendix 1)

B

Bacillus A genus of rod-shaped, Gram-positive, spore-forming bacteria which are widespread in nature. Some members of the genus are thermophiles, e.g. *Bacillus stearothermophilus* and can grow at 60 °C, others are pathogenic for Man and animals, e.g. *Bacillus anthracis* the causative agent of anthrax. Many *Bacillus* species secrete large amounts of extracellular enzymes which, in their natural habitat, allow them to degrade high-molecular-weight substrates such as starch and protein. *Bacillus* species are of great interest in biotechnology as several are used for the production of enzymes of industrial importance and other species produce antibiotics which can be used clinically.

Bacillus subtilis A Gram-positive, spore-forming, rod-shaped bacterium. The genetics and physiology of *B. subtilis* have been extensively studied and this organism has become widely established as a vehicle for genetic engineering. There are many cloning vectors available, both phage and plasmid, which will replicate in *B. subtilis*. It is non-pathogenic for Man and animals and can be consumed in large quantities with no ill effects. This, together with the fact that *Bacillus* species excrete considerable amounts of extracellular proteins, led to the organism becoming one of the main host systems for the cloning of foreign genes.

bacterial alkaline phosphatase *See* BAP.

bacteriocin A toxin or antibiotic, produced by one class of bacteria, which kills other, usually closely related, bacteria. (*See also* colicin, Col factor, Col E1)

bacteriophage A virus which infects bacteria. Commonly known as a phage. (*See* lysogeny, lytic infection, plaque, T4)

BAL-31 A nuclease from the bacterium *Alteromonas espejiana* BAL-31. One of its three activities is the simultaneous degradation of the 3′ and 5′ termini of duplex DNA. BAL-31 is used to create deletion mutants *in vitro* by progressively shortening restriction fragments from both ends. The shortened molecules can then be religated with T4 DNA ligase.

Bam **HI** (pronounced bam to rhyme with ham) A type II restriction enzyme from the bacterium *Bacillus amyloliquefaciens* H which recognises the DNA sequence below and cuts at the sites indicated by the arrows:

↓
5′ G G A T C C 3′
3′ C C T A G G 5′
↑

The four base pair sticky ends are complementary to those produced by the enzymes *Sau* 3A, *Bgl* II, *Xho* II, *Mbo* I and *Bcl* I and so fragments produced by any of these enzymes can be cloned into a *Bam* HI site. The popular cloning vector pBR322 has a single *Bam* HI site in the tetracycline resistance gene. Many vectors are constructed with a *Bam* HI site to facilitate cloning of a wide variety of DNA fragments generated by the enzymes listed above. (A full list of restriction enzymes can be found in Appendix 1)

banjo A descriptive term for a stem-loop structure in a nucleic acid molecule.

bank, gene bank A collection of recombinant DNA molecules containing inserts which together comprise the entire genome of an organism.
Also used as a verb, as in 'We'll bank *Aspergillus* in YIp5 and test for *ARS* activity in yeast.'

BAP (pronounced bap, to rhyme with tap) Bacterial alkaline phosphatase. An enzyme, isolated from *E. coli*, which removes 5′ terminal phosphate groups from DNA chains. It is used to prevent the recircularization of vector molecules during gene cloning experiments.

base The heterocyclic compounds which are the constituents of all nucleic acids. There are five common bases. Three, adenine, guanine

Fig. 3. Base structures.

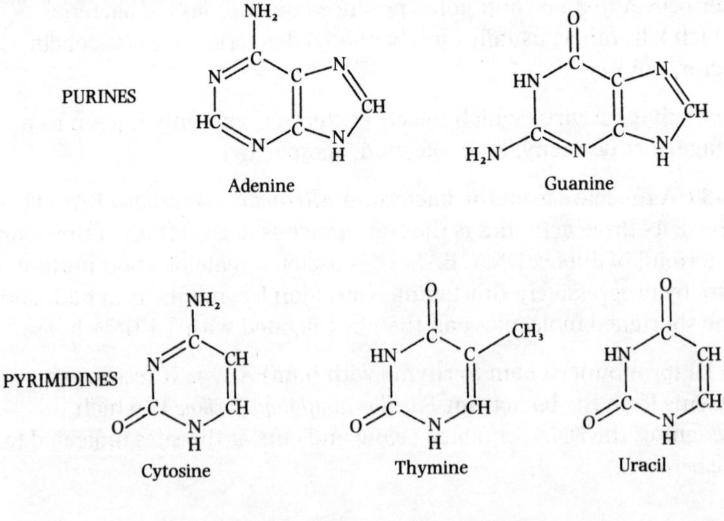

PURINES   Adenine   Guanine

PYRIMIDINES   Cytosine   Thymine   Uracil

and cytosine, are found in both DNA and RNA; thymine is found only in DNA and uracil only in RNA. A base plus a sugar (deoxyribose in DNA, ribose in RNA) is referred to as a nucleoside. A base plus sugar plus phosphate(s) is a nucleotide. The structure of the five common bases is given above.

base pair A pair of nucleotides held together by hydrogen-bonding which are found in double-stranded nucleic acids. DNA contains the base pairs A=T and G≡C, while RNA contains A=U and G≡C. (The lines indicate the number of hydrogen bonds.) The size of a nucleic acid molecule is often given in terms of the number of base pairs (bp) it contains. (*See* kb)

Benton–Davis technique *See* plaque hybridisation.

Berk–Sharp mapping, S1 mapping *See* S1 nuclease.

β-galactosidase An enzyme which will cleave lactose into glucose and galactose. The most commonly used β-galactosidase gene is from the *Escherichia coli lac* operon.

β-lactamase(s) A class of enzymes that inactivate β-lactam antibiotics (the penicillins). These enzymes are either periplasmic (in Gram-negative bacteria) or extracellular (in Gram-positive bacteria). The ampicillin resistance gene of pBR322 encodes a certain type of β-lactamase.

***Bgl* II** (pronounced buggle or baygel) A type II restriction enzyme from the bacterium *Bacillus globigii*. The enzyme recognises the DNA sequence shown below and cuts at the sites indicated by the arrows:

$$\downarrow$$
5′ A G A T C T   3′
3′ T C T A G A   5′
$$\uparrow$$

The sticky ends produced by *Bgl* II are complementary to the ends produced by the enzymes *Bam* HI, *Bcl* I, *Xho* II, *Mbo* I and *Sau* 3A. Thus, fragments produced by any one of these enzymes will have single-strand extensions which can anneal with the sticky ends on fragments produced by any other of the enzymes above. (A full list of restriction enzymes can be found in Appendix 1)

bifunctional vector or **plasmid** A DNA molecule able to replicate in two different organisms, e.g. in *E. coli* and yeast or *E. coli* and *Streptomyces*. These molecules are thus able to 'shuttle' between the two alternative hosts and are therefore also known as shuttle vectors.

It is usual to 'grow' DNA in *E. coli* for the genetic transformation of the alternative host.

9

Biodyne™ A form of activated nylon filter which can be used in place of nitrocellulose in the Southern blotting procedure. The advantage of biodyne membranes is that they are more robust than nitrocellulose and the same filter may be used in a large number of consecutive hybridisations.

biological containment A strategy which reduces the risks of recombinant molecules being propagated within microorganisms found in the general environment. Biological containment involves the use of vector molecules and host organisms which have been genetically disabled such that they can only survive in the peculiar conditions provided by the experimenter and which are unavailable outside the laboratory. (*See* physical containment)

biotinylated-DNA A DNA molecule labelled with biotin by incorporation of biotinylated-dUTP into a DNA molecule. It is used as a non-radioactive probe in hybridisation experiments such as Southern transfer. The detection of any hybrids uses a complex of streptavidin-biotin-horseradish peroxidase which will give a fluorescent green colour where hybrids have formed.

biotinylated-dUTP A nucleoside triphosphate with the vitamin biotin attached, via a spacer arm, to the dUTP (*see* Fig. 4). It can be incorporated into a DNA molecule by nick translation and then the biotinylated-DNA formed from such a reaction can be used as a probe in a hybridisation experiment. It is a non-radioactive alternative to labelling with ^{32}P.

Fig. 4. Biotinylated d-UTP.

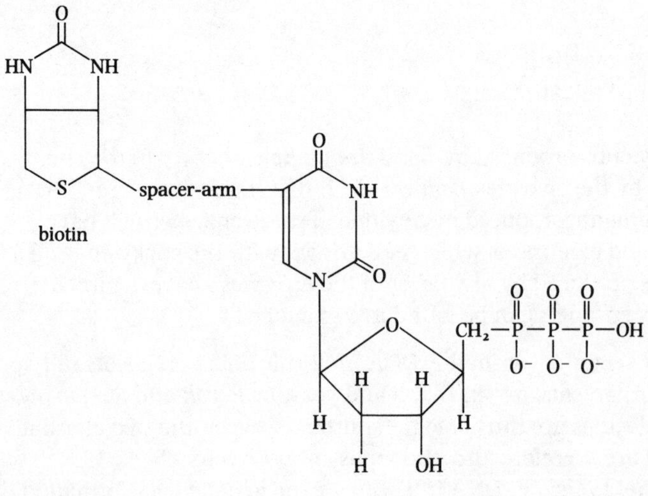

Birnboim–Doly procedure A rapid method for the purification of plasmid DNA, often used to screen recombinant colonies to determine the size of a DNA fragment inserted into a vector. The technique uses an alkaline denaturation step followed by a rapid renaturation to achieve the removal of the majority of the chromosomal DNA and most of the cell's RNA. It can be used for mini-preps (*q.v.*) or scaled up to give large quantities of fairly pure plasmid DNA. The DNA produced at the end of the procedure is usually pure enough to be digested by a restriction enzyme.

blot (i) As a verb, this means to transfer DNA, RNA or protein to an immobilising matrix such as DBM-paper, nitrocellulose or biodyne membranes. (ii) As a noun, it usually refers to the autoradiograph produced during the Southern or Northern blotting procedure. As in 'This blot demonstrates that the transforming DNA has been inserted into the chromosome.' (*See* Southern, Northern, Western blots)

blunt ends, flush ends Certain restriction enzymes, e.g. *Hae* III, generate DNA fragments which are perfectly base-paired along their entire length. The ends of such molecules are known as blunt or flush ends since they do not carry single-stranded extensions. Blunt ends may be generated artificially by removing single-stranded extensions with S1 nuclease. Blunt-end ligation is the process of joining two DNA molecules with blunt ends using DNA ligase. The process requires higher concentrations of both DNA and ligase than does the ligation of molecules with cohesive ends. The process is known colloquially as 'blunt-ending' and the term is also used as a verb as in: 'We trimmed off the 3′ extensions and blunt-ended the fragment into the *Hae* III site on our vector.'

bovine papilloma virus, bpv A group of viruses which cause warts (papillomas) in cattle and which will replicate in a wide variety of mammalian cells. These viruses do not lyse their hosts but replicate as plasmids with a copy number of 10–200 per cell. This stable association with their host has led to derivatives of bpv being used as cloning vectors for mammalian cells.

bp An abbreviation for base pair when used as a measure of the size of a double-stranded nucleic acid.

bpv *See* bovine papilloma virus.

broad host range A term used to describe a plasmid or phage which can replicate in a large number of different species.

B. subtilis *See Bacillus subtilis*.

buffer A solution containing a mixture of a weak acid and a base which resists changes in pH and is therefore able to provide a favourable environment for enzymic reactions.

buoyant density The intrinsic density which a molecule, virus or subcellular particle has when suspended in an aqueous solution of a salt such as CsCl, or a sugar, such as sucrose. The buoyant density of DNA is *ca.* 1.7 g cm^{-2} and different species have a characteristic buoyant density which reflects the proportion of G·C base pairs in the molecule. The greater the proportion of G·C the greater the buoyant density of the DNA molecule. (*See* density gradient centrifugation)

C

calf intestinal alkaline phosphatase (CIAP, sometimes pronounced 'chap' or 'sip') An enzyme which removes 5' terminal phosphate groups from DNA molecules (*see* BAP). It has the advantage over the equivalent bacterial enzyme (BAP) that it can be inactivated by heat treatment at 70 °C.

cAMP *See* cyclic AMP.

CaMV *See* cauliflower mosaic virus.

canonical sequence A prescribed or archetypical nucleotide or amino acid sequence to which all variants are compared.

cap A structure found at the 5' end of eukaryotic mRNA molecules. It consists of the modified base 7-methylguanosine joined in the opposite orientation, i.e. 5' to 5' instead of 5' to 3', to the rest of the molecule via three phosphate groups:

$$\text{m}^7\text{G}(5')\text{ppp}(5')\text{Nmp}....$$

The 7-methylguanosine cap is added to the primary transcript in the nucleus while it is being spliced and polyadenylated.

CAP, catabolite activator protein A protein studied mainly in *Escherichia coli* but with analogues in other bacteria. CAP binds cyclic AMP (cAMP) and then activates transcription from a large set of genes and operons involved in the catabolism of carbon compounds. These genes such as the *lac* operon and the *mal T* gene are thus under the positive control of the CAP–cAMP complex. The CAP–cAMP complex appears to stimulate the initial binding of RNA polymerase to the promoter. RNA polymerase has a low affinity for a CAP-activated promoter in the absence of the CAP–cAMP complex and hence the cellular concentration of cAMP controls the expression of such genes.

CAP binding site The nucleotide sequence upstream of the coding sequence of a bacterial gene or operon to which the catabolite activator protein can bind. This site only occurs before genes or operons which are under the positive control of this protein. In the presence of cyclic AMP the catabolite activator protein stimulates transcription from promoters which have a CAP binding site.

cap site The probable transcription initiation site of a eukaryotic gene. The cap is added to the 5' end of the mRNA molecule; most eukaryotic mRNAs have an A as the first nucleotide and the cap is added to that.

CAT box, CAAT box A conserved sequence found within the promoter region of the protein-encoding genes of many eukaryotic organisms. It has the canonical sequence GGPyCAATCT and is believed to determine the efficiency of transcription from the promoter.

Cauliflower mosaic virus An insect-transmitted virus which infects cauliflowers and other members of the Cruciferae. CaMV has a circular double-stranded DNA genome of *ca*. 8 kb, which contains three single-stranded 'gaps'. Viral gene products can account for 5% of the protein of the infected plant cells. CaMV is regarded as a potential vector for plant genetic engineering.

cDNA, complementary DNA The DNA complement of an RNA sequence. It is synthesised by the enzyme RNA-primed DNA polymerase or reverse transcriptase. The single-stranded DNA product of this enzyme (the reverse transcript) may be converted into the double-stranded form by DNA-primed DNA polymerase (DNA polymerase) and inserted into a suitable vector to make a cDNA clone. cDNA cloning is commonly used to achieve the expression of mammalian genes in bacteria or yeast. Since the cDNA clone is a copy of the mature mRNA molecule it contains no introns to act as a barrier to expression.

cDNA cloning A method of cloning the coding sequence of a gene starting with its mRNA transcript. It is normally used to clone a DNA copy of a eukaryotic mRNA. The cDNA copy, being a copy of a mature messenger molecule, will not contain any intron sequences and may be readily expressed in any host organism if attached to a suitable promoter sequence within the cloning vector. The procedure is as shown below (Fig. 5).

cell-free translation, cell-free system *See in vitro* translation.

cell line A clone of mammalian cells. Cell lines derived from tumour cells may be subcultured indefinitely. Those derived from normal cells, on the other hand, undergo clonal senescence; the number of doublings which they are able to go through in culture being inversely proportional to the age of the donor.

cellulose nitrate An alternative name for nitrocellulose.

CEN A cloned eukaryotic centromere. Each *CEN* is given a number corresponding to the chromosome from which it was derived. Thus *CEN* 3 refers to a cloned copy of the centromere of chromosome III. This notation is confined to yeast molecular biology at the moment but, in principle, could be widely applied. (*See* YCp)

Fig. 5. cDNA cloning.

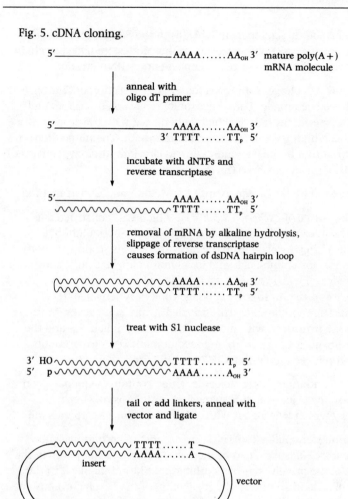

cloned cDNA copy of the eukaryotic mRNA

centromere The region of a eukaryotic chromosome which is attached to the spindle during nuclear division. It is defined genetically as that region of the chromosome which always segregates at the first division of meiosis. That is to say, it is the region of the chromosome in which no crossing-over occurs. (*See also CEN*)

chain terminator This term has two meanings: (i) Codons which do not code for an amino acid. These codons signal ribosomes to terminate protein synthesis. The codons are UAA, UAG and UGA and have been termed ochre, amber and opal respectively. Also known as stop codons or termination codons. Often two of these codons are found together at the end of a coding sequence of RNA. (ii) In the Sanger method of DNA sequencing, dideoxynucleoside triphosphates are added as chain terminators in the synthesis of a complementary DNA strand.

Charon vector A λ phage cloning vector. A large selection of Charon vectors have been made. They are usually replacement vectors and some have special mutations which, when used with the correct host strain, give a high level of biological containment. The name Charon is derived from that of the ferryman who, in Greek mythology, conveyed the spirits of the dead across the river Styx.

chemical method (of DNA sequencing) *See* Maxam and Gilbert method.

χ1776 A derivative of *Escherichia coli* K12 which was constructed as a disabled host for the cloning of DNA sequences which might be dangerous if they escaped from the laboratory. This strain has many requirements for amino acids and for diaminopimelic acid, a substance required for cell wall synthesis, and not found in the human gut. In addition, it is resistant to several phages and very sensitive to detergents such as bile salts. This crippled strain is, however, hard to grow and to transform with plasmid DNA. These difficulties and the fact that the guidelines controlling recombinant DNA experiments have been relaxed mean that χ1776 is little used.

chimera A recombinant DNA molecule which contains sequences from more than one organism. It is named after the beast in Greek mythology which was a goat with a lion's head and a serpent's tail.

chloramphenicol amplification The copy number of certain plasmids, relaxed control plasmids, can be increased by incubating *E. coli* in the presence of the protein-synthesis inhibitor, chloramphenicol. In the absence of protein synthesis, the replication of the *E. coli* chromosome ceases but that of the plasmid continues. This is a useful method for increasing the proportion of plasmid DNA in extracts of *E. coli* cells.

chloramphenicol resistance Resistance to the lethal effects of the antibiotic chloramphenicol. The most widely used gene encoding chloramphenicol resistance is from a transposon which was originally found on a large plasmid. The gene encodes a chloramphenicol acetyl transferase and has been incorporated into several cloning vectors.

chromosome A self-replicating nucleic acid molecule containing a number of genes. In bacteria, the entire genome is contained within one double-stranded, circular DNA chromosome. In eukaryotes, chromosomes are linear DNA duplexes and most organisms have their genomes divided between a number of such chromosomes; that number being characteristic for a particular species.

chromosome walking A method used to identify which clone in a gene bank contains a desired gene or sequence which cannot be selected for easily. The gene bank must contain the entire DNA sequence of the chromosome as a series of overlapping fragments. These fragments can be generated either by random shear or by partial digestion with a four base pair cutter such as *Sau*3A. A series of colony hybridisations is then carried out, starting with some cloned gene which has already been identified and which is known to be on the same chromosome as the desired gene. This identified gene is used as a probe to pick out clones containing adjacent sequences. These are then used as probes themselves to identify clones carrying sequences adjacent to them and so on. At each round of hybridisation one 'walks' further along the chromosome from the identified gene.

The diagram below illustrates how, given a clone of gene *S*, it is possible to 'walk' along the chromosomal sequence *A* through *Z* to gene *N*.

Chromosome sequence:

A B C D E F G H I J K L M N O P Q R S T U V W X Y Z

Gene bank contains the following cloned sequences:

ABC DEF GHI JKL MNO PQR STU VWX YZ..
BCD EFG HIJ KLM NOP QRS TUV WXY Z..
CDE FGH IJK LMN OPQ RST UVW XYZ

Gene *S* is used as hybridisation probe. The following clones will be identified: *STU, QRS, RST*
QRS is chosen as the probe for the next round. The following clones will be identified: *STU, QRS, RST, PQR, OPQ*
STU, QRS and *RST* are discarded as already identified. *PQR* is chosen as the probe for the next round. The following clones will be identified: *PQR, QRS, NOP, OPQ, RST*
PQR, QRS, OPQ and *RST* are discarded as already identified
NOP contains the desired gene *N*

circularization A DNA fragment generated by digestion with a single restriction endonuclease will have complementary 5′ and 3′ extensions (sticky ends). If these ends are annealed and ligated the DNA fragment will have been converted to a covalently-closed circle or circularized. (Fig. 6)

cleared lysate The intracellular contents of any cells which have been broken, treated with detergents and then centrifuged to remove large particles. The resulting (usually) clear solution contains mainly nucleic acids and protein. The preparation of a cleared lysate is usually the first step in the purification of plasmid DNA.

Fig. 6. Circularization.

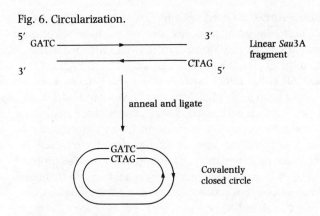

cleave This term has essentially its normal English usage and means to make a double-stranded cut in DNA with a restriction endonuclease.

clone This term is used in a number of senses. As a noun it may mean (i) a population of recombinant DNA molecules all carrying the same inserted sequence, or (ii) a population of cells or organisms of identical genotype. It is most frequently used to describe a colony of microorganisms which harbour a specific DNA fragment inserted into a vector molecule.

As a verb 'to clone' means to use *in vitro* recombination techniques to insert a particular gene or other DNA sequence into a vector molecule.

Cm^r, Cm^s Chloramphenicol resistant, chloramphenicol sensitive.

coding capacity The amount of protein which can be specified by a given DNA or RNA sequence. Estimates of coding capacity usually require assumptions about the absence of introns or of overlapping genes; as a rough guide 1 megadalton of double-stranded DNA can encode 60–70 000 daltons of protein.

coding sequence That portion of a gene which directly specifies the amino acid sequence of its protein product. Non-coding sequences of genes include control regions, such as promoters, operators and terminators as well as the intron sequences of certain eukaryotic genes.

coding strand The strand of duplex DNA which is transcribed into a complementary mRNA molecule.

<pre>
                        coding strand
        3' TACTTTCGCAAATCACCCGCGGGCATA 5'
DNA
        5' ATGAAAGCGTTTAGTGGGCGCCCGTAT 3'
mRNA   5' AUGAAAGCGUUUAGUGGGCGCCCGUAU 3'
</pre>

codon The set of three bases in an mRNA molecule which specifies an amino acid. (*See* anticodon, genetic code)

codon bias While the four constituent nucleotides of RNA can be arranged into 64 (4^3) different three-base codons, there are only twenty amino acids to be specified. This means that the genetic code exhibits considerable redundancy and a given amino acid can be specified by more than one codon. When an organism frequently uses a particular codon (rather than its alternatives) to specify a given amino acid, it is said to exhibit codon bias. Different organisms exhibit different codon biases and the phenomenon therefore represents a significant barrier to the efficient expression of cloned genes in a heterologous organism.

cohesive ends *See* sticky ends.

Col E1 A small, multicopy, colicinogenic plasmid found in *E. coli*. Vectors, such as pBR322, which use a Col E1-type replicon have a copy number of *ca.* 50 in exponentially growing cells; this can be boosted to some 3000 by chloramphenicol amplification.
   Cells carrying Col E1 synthesise and excrete large amounts of colicin when treated with low levels of a mutagen. The plasmid also encodes the *imm* protein which renders the host immune to the colicin synthesised. Other Col E1 genes encode the *mob* proteins which permit the mobilisation of the plasmid from one cell to another.

Col-factor A plasmid encoding colicin production.

colicin A protein toxin, produced by coliform bacteria such as *E. coli*, which is lethal to other bacteria. Production of, and immunity to, a colicin is usually encoded by genes on a plasmid known as a Col factor. Col factors form the basis of a number of popular cloning vectors, e.g. pBR322. (*See* bacteriocins)

colony hybridisation, Grunstein–Hogness procedure The hybridisation of a radioactively labelled probe with denatured DNA from lysed bacterial colonies. Colonies are replicated from a master plate onto a nitrocellulose filter laid on a second agar plate. After the colonies have grown up on the filter they are lysed with alkali, which also serves to denature the DNA. The filters are neutralised and the denatured DNA baked on. Radioactive DNA or RNA probe molecules are then hybridised to the DNA bound to the filter. Autoradiography subsequently reveals which bacterial colonies contain DNA complementary to the probe. The desired clones can then be retrieved from the master plate.
   Colony hybridisation may also be performed with yeast, in which case the cells in the colony must be converted to protoplasts by

treatment with a wall-lytic enzyme prior to the alkaline lysis step. The technique is named after its American inventors.

comb The plastic template used to form the slots in agarose or polyacrylamide slab gels into which the nucleic acid or protein samples are introduced.

competent A cell is said to be competent when it is able to take up nucleic acid molecules. Competent cells can occur naturally at a certain stage of growth (as in *Bacillus subtilis*), or can be induced by calcium ions at 0 °C (as in *E. coli*) or lithium ions (as in *Saccharomyces cerevisiae*). Competent cells may be frozen at − 80 °C if glycerol is present to preserve viability and are stable for several years.

complementary A nucleic acid sequence is said to be complementary to another if it is able to form a perfectly hydrogen-bonded duplex with it, according to the Watson–Crick rules of base pairing. An mRNA molecule is complementary to one of the DNA strands of the gene which encodes it.

The RNA sequence:
AUGGCAUUUCGGCCCCACUGA
is complementary to the DNA sequence:
TACCGTAAAGCCGGGGTGACT

complementary DNA *See* cDNA.

complementation The process by which the two parental genomes in a diploid cell each supply functions which the other lacks, e.g. if the two parents have the following genotypes:

Parent 1: *ADE1 leu2* – able to grow in the absence of adenine but requiring leucine in its growth medium.
Parent 2: *ade1 LEU2* – requiring the addition of adenine but not leucine for growth.
Then the diploid would have the genotype:
ADE1 leu2
ade1 LEU2
and would be able to grow without the addition of either nutrient to the culture medium.

complete digest The treatment of a DNA preparation with a restriction endonuclease for sufficient time that all of the potential target sites within that DNA have been cleaved. (*cf.* partial digest)

concatemer A long DNA molecule made up from repeated monomers of a single kind to give a linear multimer with all the monomers in the same relative orientation to one another. Phage λ replicates, at a certain stage in its life cycle, by a rolling circle mechanism to give long multimers of λ phage DNA (Fig. 7).

Fig. 7. Concatemer.

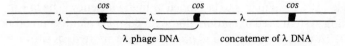

λ phage DNA   concatemer of λ DNA

concatemeric DNA is the substrate for the *in vitro* packaging of recombinant molecules constructed in cosmid or λ phage cloning experiments. In such experiments, concatemer formation is favoured by high concentrations of both DNA and ligase.

cI (c one) A repressor gene from bacteriophage λ which controls transcription from the promoters P_L and P_R. A particular mutant of this gene, cI857, produces a temperature-sensitive repressor which is inactive at 42 °C, thus enabling transcription from the strong promoters P_L and P_R to take place at this temperature. The expression of cloned genes transcribed from P_L or P_R is often controlled in this manner.

conjugation The transfer of a bacterial plasmid from one cell to another. The plasmid usually encodes the majority of the functions necessary for its own transfer.

consensus sequence If a particular nucleotide sequence is always found, with only minor variations, in a given class of genetic elements, e.g. promoters, then the usual form of that sequence is deemed the consensus sequence. Consensus sequences may also be established for genes which encode the same protein in different organisms.

constitutive An organism is said to be constitutive for the production of an enzyme or other protein if that protein is always produced by the cells under all physiological conditions. (*cf.* inducible)

containment The precautions adopted to prevent the replication of recombinant DNA molecules or genetically engineered organisms outside the laboratory. (*See* biological containment, physical containment)

copy number The number of molecules, per genome, of a plasmid or gene which a cell contains.

COS cells; COS1, COS3, COS7 These are cell lines derived from monkey cells. COS cells contain an integrated segment of SV40 DNA which codes for T antigen and so will support replication of vector molecules which contain the SV40 origin of replication, but no other DNA sequences from that virus.

21

cosmid A plasmid vector which contains the *cos* site of phage λ and one or more selectable markers such as an antibiotic resistance gene. Cosmids exploit certain properties of phage λ to enable large, 25–35 kb, DNA fragments to be cloned at high efficiency. The *in vitro* packaging system used in conjunction with cosmids requires no other phage DNA than the *cos* site, and provided there are two *cos* sites separated by *ca.* 47 kb, the DNA between the *cos* sites will be packaged. Cosmids are usually *ca.* 10 kb, and size-fractionated insert DNA of *ca.* 30 kb is used in a typical cloning experiment. The cosmid and insert DNA are ligated at high DNA and ligase concentrations to produce concatemers which are substrates for the packaging reaction. Cosmids and cosmid recombinants replicate as plasmids (Fig. 8).

Fig. 8. Cosmid cloning.

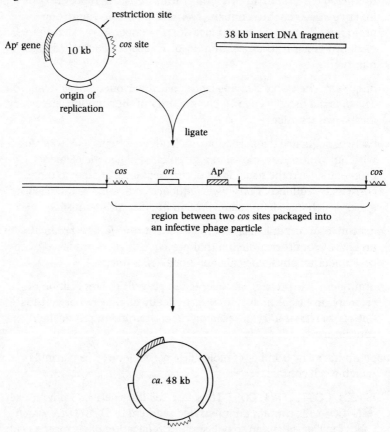

region between two *cos* sites packaged into an infective phage particle

Infection of *E. coli* yields ampicillin-resistant transductants containing recombinant cosmid molecule which replicates as a plasmid

cos **site** These are the cohesive ends of certain phage DNA molecules. The *cos* site usually referred to is from λ. λ DNA, in the viral particle, is a double-stranded linear molecule of 49 kb with 12 bp cohesive or sticky ends. These ends are formed during packaging of concatemeric λ DNA into viral head particles. The λ-*ter* gene product recognises the double-stranded 12 bp *cos* sequence and cuts it asymmetrically to generate the sticky ends. Plasmids containing the *cos* site of λ are called cosmids.

C_0t (pronounced COT) Original concentration (of DNA) × time. It is the parameter used in the renaturation analysis of DNA genomes. Highly reiterated sequences will renature at low C_0t values while unique sequences will renature at high C_0t.

co-transduction Transducing phages carry some host DNA either in place of the phage genome or integrated within it. Two host genes are said to undergo co-transduction if they are both transferred within the same phage particle. Since the amount of DNA carried in a phage head is small, this implies that the two genes are close together on the host chromosome. The study of co-transduction frequencies is, therefore, a useful method for mapping bacterial genes.

co-transformation In genetic engineering experiments it is often necessary to transform with a plasmid for which there is no selectable phenotype and then screen for the presence of that plasmid within the host cell. Since not all host cells take up DNA (are competent) this may be very laborious. Co-transformation is a technique in which host cells are incubated with two types of plasmid, one of which is selectable and the other not. Cells which have been transformed with the first plasmid are then selected. If transformation has been carried out at high DNA concentration then it is probable that these cells will also have been transformed with the second (non-selectable) plasmid. The technique is frequently used in experiments with mammalian cells.

covalently closed circle, CCC A double-stranded DNA molecule with no free ends. The two strands are interlinked and will remain together even after denaturation. In its native form a CCC will adopt a supercoiled configuration. (*See* ethidium bromide, intercalating agent, open circle, plasmid and Fig. 9)

cross-over The site of the reciprocal exchange of genetic material during an *in vivo* recombination event. The cross-over is the place where the breakage and reunion of DNA strands take place. (Fig. 10)

crown-gall, crown-gall disease A tumour formed, usually, on the stems of broad-leaved plants when infected with *Agrobacterium tumefaciens*

23

Fig. 9. Covalently closed circles.

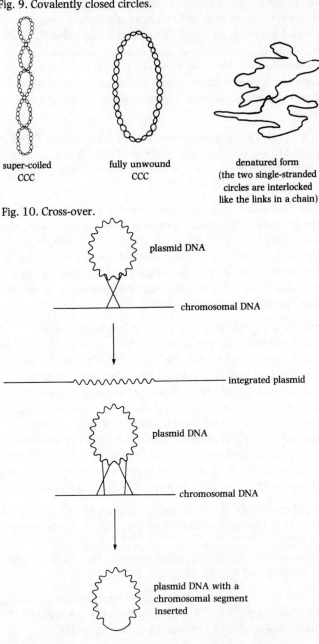

super-coiled   fully unwound   denatured form
CCC   CCC   (the two single-stranded
                                             circles are interlocked
                                             like the links in a chain)

Fig. 10. Cross-over.

plasmid DNA

chromosomal DNA

integrated plasmid

plasmid DNA

chromosomal DNA

plasmid DNA with a
chromosomal segment
inserted

chromosomal DNA
with a plasmid
segment inserted

24

containing a Ti-plasmid. The bacterium is only necessary for the initiation of the tumour. The genome of the affected plant cells contains several copies of a segment of the Ti-plasmid (the T-DNA). Crown galls can be of two types, octopine or nopaline, depending on the type of Ti-plasmid which initiated the tumour. Whole plants can be regenerated from crown-gall tissue and some of these still contain the T-DNA.

CRP, cAMP receptor protein An alternative name for CAP.

cryptic plasmid A plasmid whose presence causes no discernible change in the phenotype of the host.

CsCl – caesium (cesium) chloride A salt which forms very dense solutions in water or aqueous buffers. It is used in isopycnic centrifugation to separate DNA molecules of different densities. When spun at high speed in an ultracentrifuge Cs^+ atoms distribute themselves in a density gradient with a useful density range ($ca.$ 1.5–1.8 g cm^{-2}) for the separation of DNA molecules.

Cs_2SO_4 – caesium (cesium) sulphate Like CsCl this caesium salt forms dense solutions which give density gradients on centrifugation. Cs_2SO_4 gradients reach densities greater than those of CsCl and therefore are used in the isopycnic centrifugation of double-stranded RNA molecules.

curing The elimination of a plasmid from its host cell. Many agents which interfere with DNA replication, e.g. ethidium bromide, can cure plasmids from either bacterial or eukaryotic cells.

cut Slang term meaning to make a double-stranded break in DNA, usually with a type II restriction endonuclease. E.g. 'The DNA was cut with EcoR1 and run out on a 1% agarose gel.' ($cf.$ nick)

cyclic AMP, cAMP This compound is derived from ATP by the action of the enzyme adenyl cyclase. cAMP is an important regulatory molecule; in higher eukaryotes it is a mediator of hormone action, while in bacteria it is involved in the catabolite repression of gene expression. (*See* CAP, CAP-binding site)

cycloserine enrichment A method of enriching a bacterial population for those members which cannot divide in a given medium. It is based on the fact that cycloserine will only kill dividing cells. It is often used to enrich a population of cells for those which have a piece of foreign DNA inserted into the tetracycline-resistance gene of a cloning vector. Tetracycline-sensitive cells will not divide in a medium containing tetracycline, a bacteriostatic antibiotic, while tetracycline-resistant cells will grow and be killed by the cycloserine. The cells are harvested by centrifugation and put in fresh, drug-free medium. This procedure is repeated several times to enrich the population for tetracycline-sensitive (recombinant) cells.

D

dA Deoxyadenosine, one of the four nucleoside constituents of DNA. For structure, *see* base.

dA.dT tailing *See* A's and T's method.

dalton A unit of molecular weight (roughly equivalent to the mass of a hydrogen atom). Molecular weight is a ratio and therefore has no units but molecular biologists commonly use the dalton (named after the famous nineteenth-century chemist, John Dalton) in shorthand expressions of molecular weight, especially as megadaltons. One megadalton is equal to 1×10^6 daltons.

dam⁻ An *E. coli* mutant which is deficient in the methylation of its own DNA. The methylation of DNA in *dam*⁺ strains modifies the DNA so that it is no longer recognised by certain restriction enzymes, e.g. *Bcl*1.

DBM-paper, diazobenzyloxymethyl paper An activated paper used to covalently bind nucleic acids for hybridisation experiments. It is particularly useful for binding RNA in the Northern blotting procedure. It has the structure shown below.

Fig. 11. DBM-paper (structure).

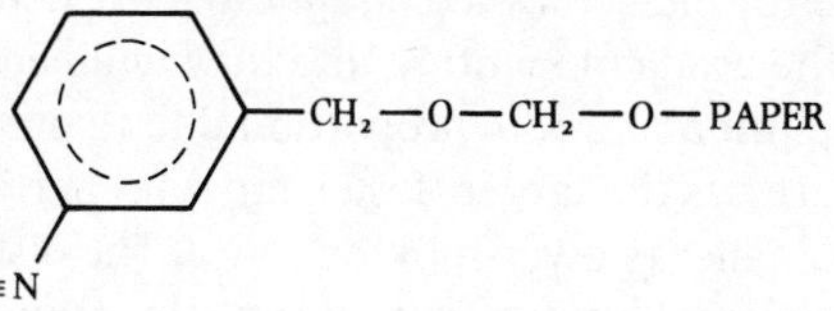

dC Deoxycytosine, one of the four nucleoside constituents of DNA. For structure, *see* base.

ddNTP *See* dideoxynucleotide.

δ, δ sequence The repeat sequences on the ends of the Ty series of yeast transposons. The nucleotide sequence of δ is highly conserved between different Ty elements. A Ty element may be excised from a chromosome by a single cross-over between its two δ sequences. The single copy of δ which remains in the chromosome after such an event is known as a solo δ.

Denhardt's solution A solution used in DNA hybridisation procedures such as the Southern blot technique. It is used to coat a nitrocellulose filter after blotting to inhibit nonspecific binding of the probe. It comprises 0.1 g Ficoll, 0.1 g polyvinylpyrrolidone and 0.1 g bovine serum albumin per 100 ml of water.

denaturation The destruction of the secondary or tertiary structure of a protein or nucleic acid by physical or chemical means. Most commonly used by genetic engineers to describe the destruction of hydrogen bonds maintaining the double-stranded nature of all or part of a DNA or RNA molecule.

density gradient centrifugation The separation of macromolecules or subcellular particles by sedimentation through a gradient of increasing density under the influence of a centrifugal force. The density gradient may either be formed before the centrifugation run by mixing two solutions of different density (as in sucrose density gradients) or it can be formed by the process of centrifugation itself (as in CsCl and Cs_2SO_4 density gradients).

Separations achieved by this technique depend on either
(i) The buoyant density of the particles when separation is by density gradient equilibrium centrifugation. Particles migrate through the gradient until they reach the place where the density of the gradient solution is equal to their own. This is the isopycnic point of the molecule or particle. The particle will not migrate from this point no matter how long the centrifugation run lasts.
or (ii) Separation by size and shape which is performed by rate zonal centrifugation. The particles or molecules to be separated are denser than any other portion of the gradient solution and they will be sedimented through that solution at a rate proportional to their size. As the centrifugation run progresses the larger (faster-moving) particles will be separated from the smaller (slower-moving) ones. The distance between different sized particles will increase with centrifugation time, but equilibrium is not reached and eventually all particles will sediment to the bottom of the centrifuge tube.

deoxyadenosine *See* dA.

deoxycytidine *See* dC.

deoxyguanosine *See* dG.

deoxynucleoside triphosphate *See* nucleotide.

deoxynucleotidyl transferase *See* terminal transferase.

deoxyribonucleoside *See* nucleoside.

deoxyribonucleotide *See* nucleotide.

deoxythymidine *See* dT.

dG Deoxyguanosine, one of the four nucleoside constituents of DNA. For structure, *see* base.

dG.dC tailing A method of cloning random fragments of DNA by the addition of complementary homopolymer tails to the insert and the vector DNAs. The details are given under A's and T's method; the technique of dG·dC tailing is the same except that oligo dG and oligo dC tails are added by terminal transferase instead of oligo dA and oligo dT. dG·dC tailing has the advantage that the GC base pairs produce more stable hybrid molecules than AT base pairs. More importantly, if the vector molecule is cut at a *Pst*1 site and tailed with Gs and the foreign DNA fragment is tailed with Cs, then the *Pst*1 site is reconstructed in the recombinant molecule, thus permitting easy retrieval of the cloned fragment.

DHFR, dihydrofolate reductase An enzyme involved in the synthesis of purines and pyrimidines. It is inhibited by trimethoprim in prokaryotes or by methotrexate in prokaryotes and eukaryotes. Genes for certain methotrexate-resistant DHFRs can be used as selective markers on cloning vectors for prokaryotic and eukaryotic cells.

dideoxynucleoside triphosphate, ddNTP A nucleoside triphosphate which lacks a 3′ hydroxyl on the deoxyribose moiety. It can be incorporated into a DNA chain by, for example, the Klenow enzyme but since it lacks a 3′ hydroxyl for the next nucleoside triphosphate to be joined to, the growing DNA chain will stop there. ddNTPs are used in the Sanger method of DNA sequencing.

Fig. 12. A 2′,3′ dideoxynucleoside triphosphate (structure).

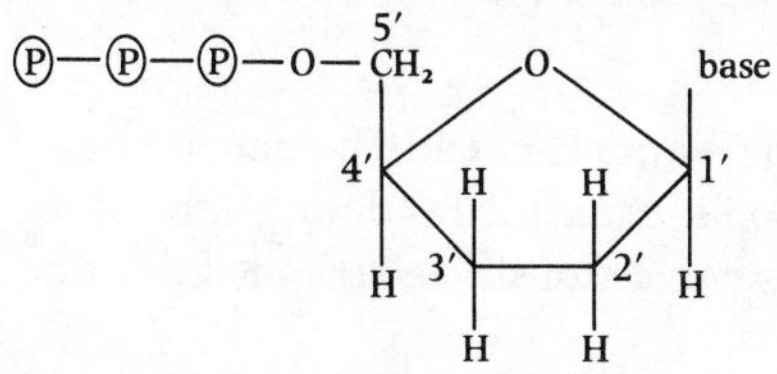

dideoxysequencing *See* Sanger method.

differential centrifugation A method of separating subcellular particles according to their sedimentation coefficients which are roughly proportional to their size. Cell extracts are subjected to a succession of centrifuge runs at progressively faster rotation speeds. Large particles, such as nuclei or mitochondria, will be pelleted at relatively slow speeds, while higher *g* forces will be required to sediment small particles such as ribosomes.

digestion The treatment of a substrate molecule with an enzyme preparation in which covalent bonds are hydrolysed.

dihydrofolate reductase *See* DHFR.

directed mutagenesis *See* site-specific mutagenesis.

direct repeat Two or more stretches of DNA within a single molecule which have the same nucleotide sequence in the same orientation. Direct repeats may be either adjacent to one another or far apart on the same molecule.

TATTA....TATTA
ATAAT....ATAAT is an example of a direct repeat.

(*cf.* inverted repeat)

disc gel A gel, usually of agarose or polyacrylamide, cast in a vertical tube instead of a slab. The nucleic acid or protein bands are resolved as a series of discs in the cylindrical gel. The method is little used now, mainly because of the difficulty of ensuring that each individual gel is run in an identical manner. However, disc gels, and their variants, still find application in preparative procedures.

D-loop, displaced loop, displacement loop A D-loop can form when a supercoiled DNA molecule is incubated, in the correct conditions, with a short, single-stranded DNA fragment which is homologous to some part of the supercoiled molecule. A strand of the supercoiled molecule is displaced by the short DNA fragment to form a D-loop which is then available for manipulation. This is the initial step in one form of site-specific mutagenesis and in the replication of many circular DNA molecules *in vivo*. (*See* R-looping)

DNA, deoxyribonucleic acid, desoxyribonucleic acid The genetic material of all organisms and organelles so far examined is double-stranded DNA. A number of viral genomes consist of single-stranded DNA or single- or double-stranded RNA.

In double-stranded DNA the two strands run in opposite directions (are antiparallel) and are coiled round one another in a double helix. Purine bases on one strand specifically hydrogen bond with pyrimidine bases on the other strand according to the Watson–Crick rules (A pairs with T, G pairs with C). This maintains a constant width for the double helix of 20 Å (2.0 nm). In the B-form, DNA adopts a right-hand helical conformation, with each chain making a complete turn every 34 Å (3.4 nm) or once every ten bases. (Fig. 31)

Fig. 13. DNA.

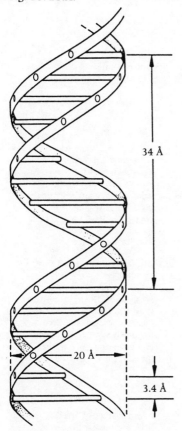

DNA ligase *See* ligase.

DNA polymerase Any enzyme which synthesises DNA by copying a template strand. DNA polymerases synthesise DNA in the 5′ to 3′ direction by successively adding nucleotides to the free 3′ hydroxyl group of the growing strand. The template strand determines the order of addition of nucleotides via Watson–Crick base-pairing. *E. coli* has three DNA polymerases: Polymerase I (see below) is involved in repair synthesis, while DNA polymerase III is responsible for DNA replication. The function of Polymerase II is unknown at present. (*See* DNA polymerase I, nick translation, Okazaki fragment)

DNA polymerase I, Pol I *E. coli* DNA polymerase I has three enzymatic activities which enable it to repair damaged DNA within the cell: (i) 5′–3′ polymerase, (ii) 5′–3′ exonuclease and (iii) 3′–5′ exonuclease.

31

Activities (i) and (ii) are exploited in the nick-translation technique for labelling DNA *in vitro*.

Fig. 14. DNA polymerase I.

(A)

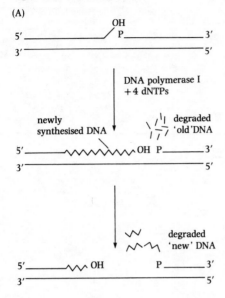

A nicked duplex is the substrate for the enzyme.

The 5'⟶3' exonuclease activity degrades the nicked strand and the 5'⟶3' polymerase activity replaces the degraded material with newly synthesised DNA. These two activities are used to label DNA to high specific activity in nick translation.

At the same time the 3'⟶5' exonuclease activity may act to degrade the newly synthesised DNA. This process is used to remove misincorporated bases from the newly synthesised strand and hence is known as the 'proof-reading' activity.

(B)

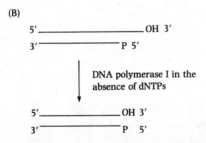

In the absence of dNTPs, DNA polymerase I is used *in vitro* to convert sticky-ended molecules into blunt-ended ones. The 3'⟶5' exonuclease activity degrades the single-stranded extension.

DNase, deoxyribonuclease Any enzyme which degrades DNA. (*See* endonuclease, exonuclease, restriction enzyme)

dot-blot A variation of the Southern blot procedure which is used to determine the concentration of a particular RNA or DNA species. Different concentrations of the non-radioactive DNA are denatured and applied as a dot to nitrocellulose paper or other DNA support matrix. This is then hybridised with the radioactive probe under study. After autoradiography the intensities of the radioactive images formed are quantified and compared to a control series to determine the concentration of the non-radioactive molecule.

double-stranded *See* duplex.

downstream A term used to describe the relative positions of sequences on a nucleic acid or protein molecule. It means, in the direction in which a nucleic acid or protein molecule is synthesised, i.e. on the 3′ side of any given site in DNA or RNA, and on the C-terminal side of any site within a polypeptide.

Note that when referring to the downstream regions of a gene in respect to its activity as a template for RNA synthesis, it is the 5′ side of any given region in the coding, or sense, strand which is denoted.

***Drosophila*, fruit fly** A dipteran fly which was a central experimental organism in the development of classical genetics. Experiments on *Drosophila* led to the chromosome theory of inheritance by Morgan, Bridges, Sturtevant and Muller. *Drosophila* continues to be an important experimental organism in developmental, behavioural and population genetics.

drug resistance *See* antibiotic resistance.

ds Double-stranded. (*See* duplex)

dT Deoxythymidine, one of the four nucleoside constituents of DNA. For structure, *see* base.

duplex A double-stranded nucleic acid molecule or a double-stranded region of a mainly single-stranded molecule.

E

Eckhardt method A rapid method for the analysis of plasmid DNA from recombinant colonies. The colonies are lysed and the DNA applied directly to an agarose gel or, sometimes, the colonies are actually lysed in the well of the gel. Plasmids containing insert DNA are larger than the recircularized vector molecule and have a reduced electrophoretic mobility which is readily detected by this method.

E. coli, Escherichia coli A Gram-negative, non-spore-forming, rod-shaped bacterium which is a normal inhabitant of the lower intestine of most mammals including Man. *E. coli* has become the most intensively studied microorganism and enough is now known about its genetics and biochemistry to make it the host of choice in the majority of cloning experiments. The wide choice of vectors and techniques for sophisticated *in vivo* genetic manipulation means that *E. coli* will probably always be used in molecular biology.

There are, however, some drawbacks to its use. Certain wild-type strains are pathogenic and virtually all strains have a lipopolysaccharide component of their outer membrane which is intensely immunogenic. This has led to other organisms being developed for use in large-scale production where the final product is to be used in domestic animals or Man.

***Eco* RI** (pronounced echo R one) A type II restriction enzyme from *E. coli* RY13. *Eco* RI recognises the DNA sequence shown below and cuts at the sites indicated by the arrows:

$$\downarrow$$
5′ G A A T T C 3′
3′ C T T A A G 5′
$$\uparrow$$

It is a widely used restriction enzyme being present in very large amounts in cells which carry the plasmid pMB1 and relatively easy to isolate. Under conditions of low Mg^{2+} ion concentration and a pH of 8.5 or by substituting Mn^{2+} for Mg^{2+} the sequence specificity of *Eco* RI reduces to

$$\downarrow \qquad\qquad \downarrow \qquad\qquad\qquad \downarrow$$
5′ N A A T T C 3′   or   G A A T T N   and   A A T T
3′ N T T A A G 5′   C T T A A N   sometimes   T T A A
$$\uparrow \qquad\qquad\qquad \uparrow \qquad\qquad\qquad\qquad \uparrow$$

This is known as the *Eco* RI* activity or simply 'star' activity. Many vectors are constructed to make use of *Eco* RI in cloning experiments, e.g. both pBR322 and pBR325 contain a single *Eco* RI site which, in the latter, is within the chloramphenicol-resistance gene so permitting screening for inserts by insertional inactivation. (A complete list of restriction enzymes can be found in Appendix 1)

electro-blotting The electrophoretic transfer of macromolecules (DNA, RNA or protein) from a gel in which they have been separated to a support matrix such as a nitrocellulose sheet. An alternative to the capillary transfer usually used in techniques such as Southern and Northern blotting.

ELISA, enzyme-linked immunosorbent assay A sensitive immunological assay system which avoids both the hazards and expense of radioactive or fluorescence detection systems. Two antibody preparations are used in ELISA. The primary antibody binds the antigen and is itself bound by the second, antiglobin, antibody. The antiglobin is linked to an enzyme, e.g. horseradish peroxidase, whose activity is easily monitored, for instance by a colour change. The extent of the enzymic reaction is then a quantitative indication of the amount of primary antibody or, indirectly, of antigen present.

Elu-tip™ *See* NACS™.

end-labelling The introduction of a radioactive atom at the end of a DNA or RNA molecule. A commonly used method is to use T4 polynucleotide kinase to introduce a ^{32}P atom onto the 5′ end.

endonuclease An enzyme which cuts within a polynucleotide chain.

enhancer element, enhancer sequence A sequence found in eukaryotes and certain eukaryotic viruses which can increase transcription from a gene when located (in either orientation) up to several kilobases from the gene being studied. These sequences usually act as enhancers when on the 5′ side (upstream) of the gene in question. However, some enhancers are active when placed on the 3′ side (downstream) of the gene. In some cases enhancer elements can activate transcription from a gene with no (known) promoter.

enrichment The process of increasing the proportion of mutant cells in a culture. Enrichment is used when the desired mutants cannot be selected for directly. The technique is essentially the converse of selection. Conditions are established such that the desired mutants will not grow. Wild-type (growing) cells are then killed by some physical or chemical treatment which does not affect non-growing (mutant) cells. Such methods include treatment with drugs such as penicillin (for bacteria) or nystatin (for fungi), inositol starvation, and heatshock.

Usually, only a proportion of wild-type cells are killed by such treatments and several rounds of enrichment may be necessary. (*See* cycloserine enrichment)

entrapment, liposome entrapment The technique of encapsulating DNA or other molecules within a liposome in order to facilitate their transfer through a cell membrane.

episome An old name for an extrachromosomal element which replicates within a cell independently of the chromosome and is able to integrate into the host chromosome. The step of integration may be governed by a variety of factors and so the term episome has lost favour and been superseded by the wider term plasmid.

Eppendorf The name of a West German scientific instrument manufacturer which has entered the language of the molecular biologist in a manner analogous to the use of 'Hoover' or 'Biro' in the English language. There are two meanings: (i) a bench-top microcentrifuge, the plastic tubes for which are called 'Eppendorf tubes', and (ii) an automatic micropipette whose disposable plastic tips are referred to as 'Eppendorf tips'. As with 'Hoover' or 'Biro', Eppendorf is used to describe the appropriate instrument irrespective of who actually manufactured it.

Escherichia coli See E. coli.

ethidium bromide, EBr, EtBr An intercalating agent which allows the ready detection of double-stranded nucleic acid molecules in agarose gels. The nucleic acid/ethidium bromide complex fluoresces brightly when exposed to ultraviolet (UV) light.

Ethidium bromide is also used in the isolation of covalently closed circular plasmid molecules in CsCl gradients. The intercalation of ethidium bromide into a DNA molecule increases the length of the DNA and reduces its density. Covalently closed circular molecules (CCCs) can bind less ethidium bromide than linear or open-circular molecules since their ability to 'stretch' in order to accommodate the intercalating agent is limited by the supercoiling of the molecule. The density of CCCs is therefore reduced to a lesser extent than linear or open circular molecules and the CCCs will appear relatively more dense in a CsCl–EBr gradient. EBr has the chemical structure shown in Fig. 15.

eukaryote, eukaryotic Organisms which have the following cellular organisation: (i) genome divided into a number of chromosomes which are contained within a nucleus and are separated from the cell cytoplasm by the nuclear membrane; (ii) gene expression in eukaryotes does not involve the organisation of genes into operons, genes may be

Fig. 15. Ethidium bromide (structure).

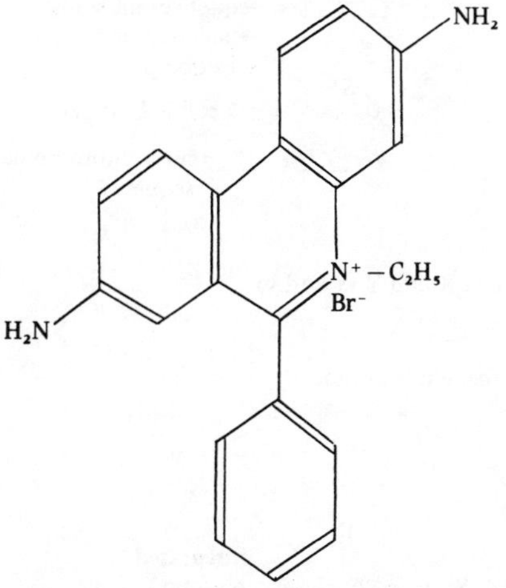

divided up by non-coding sequences called introns; (iii) there are three different RNA polymerase complexes with different transcriptional specificities; and (iv) the cytoplasm itself contains a number of membrane-bound organelles which may have their own genomes and gene expression systems, e.g. mitochondria and chloroplasts.

eviction, gene eviction A method, originally developed in yeast molecular biology, which permits the retrieval of a chromosomal copy of any gene which has previously been cloned. Thus if the wild-type copy of a gene has been cloned gene eviction may be used to obtain a mutant copy and *vice versa*. A feature of the gene eviction procedure is that DNA sequences adjacent to the gene of interest are retrieved at the same time and so the method may be used for chromosome walking.

The procedure is shown in the diagram below (Fig. 16). The host organism is transformed with an integrative plasmid (e.g. see YIp). The recombinant plasmid integrates itself into the host's chromosome by homologous recombination, thereby creating a duplication. Subsequent treatment of the transformant's DNA with a restriction enzyme generates a linear molecule which contains the bacterial sequences required for replication and selection in *E. coli* together with a hybrid region containing host and plasmid DNA from the chromosome segment originally cloned. This linear fragment will also contain some flanking sequences from the chromosome. The fragment may be circularized using DNA ligase and transformed back into *E. coli* where it will replicate.

excrete *See* export.

Fig. 16. Eviction.

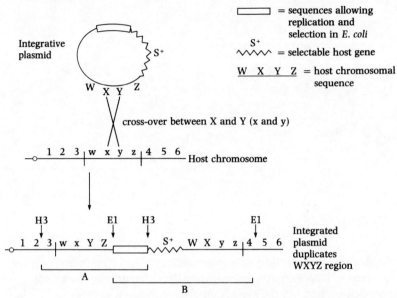

cross-over between X and Y (x and y)

Cutting with Hind III at the two sites H3 releases linear fragment A which may be circularized and transformed into *E. coli*. A contains the wx chromosomal sequences from the cloned region plus the leftward sequence 3. Cleavage at the two E1 sites by *Eco* RI generates fragment B which may also be circularized into a plasmid which can replicate and be selected in *E. coli*. In B, chromosomal sequences y and z have been evicted together with the rightward flanking sequence 4.

excretion vector A cloning vector which has the coding sequence for a signal peptide next to a restriction enzyme site into which foreign DNA can be cloned. If the foreign DNA fragment contains the coding sequence of a gene in phase with the signal peptide the protein encoded by the foreign gene will be excreted from the cell.

exon That part of a split gene which is expressed in the final protein or RNA product of that gene (*cf.* intron). (See Fig. 17.)

exonuclease An enzyme which requires a free end in order to degrade a DNA or RNA molecule (*cf.* endonuclease). 5′ exonucleases require a free 5′ end and degrade the molecule in a processive manner in the 5′ → 3′ direction. 3′ exonucleases require a free 3′ end and degrade the molecule in the opposite direction.

exonuclease III, Exo III A multifunctional enzyme from *E. coli*. The activity most widely used is the 3′ → 5′ exonuclease activity. This is

Fig. 17. Exon.

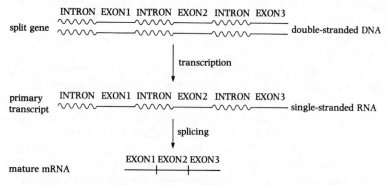

used to create long single-stranded regions in duplex DNA for subsequent sequencing by the Sanger method.

export The transport of a protein out of a cell. Alternative terms are secrete and excrete. Secretion is a general term for the transport of proteins across any membrane, either into some cellular compartment (such as a vacuole) or into the medium. Excretion, like export, implies that the protein crosses the cell membrane into the surrounding medium.

expression vector A vector containing a promoter sequence which facilitates the very efficient transcription of an inserted gene and therefore results in the cell containing a high concentration of the insert's protein product. Such promoters are described as 'high level'; examples include P_L (the leftward promoter of phage λ) and the promoter of the yeast *PGK* (phosphoglycerate kinase) gene.

extension The single-stranded DNA tail found at the end of a restriction fragment. Different restriction endonucleases generate extensions at either the 5′ or 3′ end of the DNA strand.

extrachromosomes Self-replicative genetic elements separate from the main chromosome(s) of a cell. The definition usually excludes viruses, but the division is somewhat arbitrary.

In bacteria, plasmids are the principal extrachromosomes; they encode functions which are not essential to the growth and division of the host cell. In eukaryotes, extrachromosomes may be either essential or dispensable. They may inhabit (i) the nucleus, e.g. extrachromosomal rDNA molecules, yeast 2 μ plasmid, (ii) the cytosol, e.g. dsRNA molecules in fungi, or (iii) the cytoplasmic organelles, e.g. mitochondrial DNA, chloroplast DNA. Eukaryotic extrachromosomal elements may be recognised genetically by their failure to show segregation at meiosis.

F

F, F factor, F plasmid, F sex factor A large, 94.5 kb, transmissible plasmid found in some strains of *E. coli*. The F denotes fertility since the F plasmid can not only transfer itself from one strain of *E. coli* to another but can mobilise portions of the chromosome. *E. coli* cells carrying an F plasmid (F^+, male or donor cells) will transfer the F plasmid to F^- (female or recipient) cells. The F plasmid can integrate into the host chromosome converting the *E. coli* cell from F^+ to the Hfr state. Hfr stands for high frequency recombination. In these cells the plasmid mediates not only its own transfer but also that of adjacent *E. coli* genes which are then available to recombine with the recipient genome. The F plasmid can also acquire portions of the host chromosome and it is then known as an F' plasmid. (*See* conjugation)

fd A phage which attacks *E. coli* cells which carry surface structures called pili specified by the F plasmid. The genome of the viral particle is a single-stranded DNA circle of 6408 nucleotides, while the replicative form within the cell is a double-stranded circle. fd is *ca.* 97% homologous to M13. Several cloning vectors with similar properties to the M13 Mp vectors have been constructed but are not in such widespread use.

5′ (five prime) The atoms making up the ring of a purine or pyrimidine base are numbered in turn, 1, 2, 3, etc. In order to distinguish them, the constituent atoms of the pyranose ring of the ribose or deoxyribose sugar to which the base is attached in a nucleoside are numbered 1′, 2′, 3′, etc. It is the 5′ carbon atom to which the phosphate groups are attached in nucleotides.

Fig. 18. 5′ (five prime) (nucleotide numbering system).

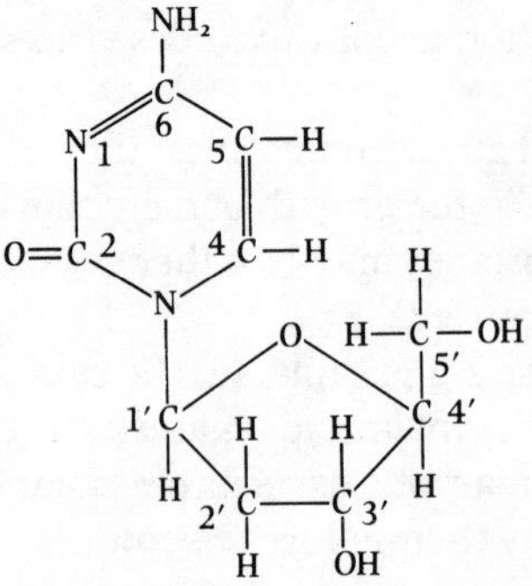

As an example of the numbering system the structure of cytosine deoxyribonucleoside (cytidine) is given above (Fig. 18). The oxygen which closes the pyranose ring is not numbered.

5′ extension The single-stranded cohesive end left at the 5′ end of a restriction fragment. For example, *Bam* HI cuts DNA as shown below. Fig. 19. 5′ extension.

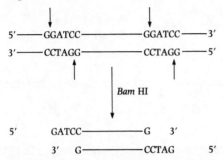

The restriction fragment generated has single-stranded extensions at its 5′ end (*cf.* 3′ extension).

flush ends *See* blunt ends.

fold-back The structure formed when a double-stranded DNA molecule containing an inverted repeat sequence is denatured and then allowed to reanneal at low DNA concentrations. The repeated sequence permits the formation of a double-stranded region within each of the separated strands of the original molecule.

foldback element, FB element A member of a family of transposons found in the genome of *Drosophila*. Foldback elements have long inverted repeats at their termini.

footprint, protection experiment A method which uses certain of the Maxam and Gilbert DNA sequencing reactions to determine the nucleotides which are close to, or in contact with, a DNA-binding protein. The protein is first bound to the DNA and then the DNA is subjected to one or more chemical modifications, or digested with DNase I. When the reaction products are run out on a sequencing gel next to a complete sequence of the DNA fragment, a characteristic loss of bands, the footprint, is seen where the protein protected the DNA from chemical modification; in some cases the bands are enhanced by the protein.

forced cloning A cloning strategy in which vector and insert DNA are cut with the same pair of restriction enzymes. The linear molecule generated will thus have non-complementary ends, so self-ligation is prevented, and recombinant molecules are the only products of circularization.

four base pair cutter A type II restriction enzyme which recognises and cuts within a four base pair sequence. E.g.

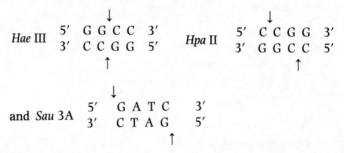

Such enzymes cut genomic DNA very frequently, once every 256 (4^4) base pairs on average. For this reason partial digests with these enzymes are often used to generate approximately random fragments of DNA for the creation of gene banks.

fragment, restriction fragment The individual polynucleotides produced by the digestion of DNA with a restriction endonuclease.

freeze-squeeze A method of recovering DNA from an agarose gel. The portion of the gel containing the desired class of molecules is cut out and frozen. It is then placed between two sheets of Parafilm™ and squeezed until a drop of DNA solution is ejected. It is claimed that *ca* 60% of the DNA contained in the gel fragment may be recovered in this way.

fusion protein If parts of two protein-encoding genes are ligated together so that their reading-frames remain in phase, then expression of the hybrid gene will result in the production of a fusion protein with its N-terminal portion encoded by the 5′ end of one gene and its C-terminal portion by the other gene. There are two reasons for fusing genes to generate fusion proteins: (i) to put the expression of some

Fig. 20. Fusion protein.

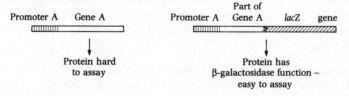

heterologous gene under the control of a strong promoter; and (ii) to facilitate the study of some gene of interest by substituting some protein with a more easily assayed function for its normal gene product. This type of fusion is usually made to the *E. coli* gene for β-galactosidase; they are known as *lacZ* fusions.

G

gap A missing section on one of the strands of a DNA duplex. The DNA will therefore have a single-stranded region.

gel The inert matrix used for the electrophoretic separation of nucleic acids or proteins.

Molecule:	DNA	RNA	Protein
Gel matrix:	agarose	agarose or polyacrylamide	polyacrylamide or (rarely) starch

gene The unit of hereditary function. A DNA or, for some viruses, RNA sequence which encodes a functional protein or RNA molecule.

gene bank A collection of cloned DNA fragments from a single genome. Ideally the bank should contain cloned representatives of all the DNA sequences in the genome. For large genomes, such banks are made by shot-gun cloning.

genetic code The set of rules which governs the relationship between the linear order of the four different nucleotides in an mRNA molecule and that of the twenty different amino acids in the protein which it encodes. Each set of three nucleotides, a triplet, specifies a single amino acid and is known as a codon. The genetic code is non-overlapping: a mutation which alters only a single nucleotide in a gene can only change one amino acid in the encoded protein and not two or three. Many mutations do not alter the amino acid sequence of the protein at all. This is because the code is degenerate with 64 (4^3) codons specifying twenty amino acids so many amino acids are determined by more than one codon. (The codon assignments of the twenty amino acids are shown in Appendix 5) (*See* initiation codon, stop codon for explanations of the punctuation of the code)

genetic engineering A popular term for the use of *in vitro* techniques to produce DNA molecules containing novel combinations of genes or other sequences.

genome The complete set of genes of an organism, organelle or virus, e.g. human (nuclear) genome, chloroplast genome, phage genome.

genomic An adjective, usually applied to purified DNA from a particular source.

genotype The genetic constitution of an organism as revealed by genetic or molecular biological analysis. E.g. a diploid organism contains two alternate alleles (forms) of a gene, X and x, with X dominant to x. Only the presence of X will be revealed in the organism's phenotype. Genetic analysis will reveal that its genotype is Xx.

G418 resistance *See* kanamycin resistance.

glyoxal, ethanedial (OHCCHO) A powerful denaturing agent which is used to denature double-stranded nucleic acid molecules, or destroy any secondary structure present in single-stranded molecules, prior to gel electrophoresis.

GMAG (pronounced 'jee-mag') Genetic Manipulation Advisory Group. The body in the United Kingdom which advised on the physical and biological containment procedures which should be employed in recombinant DNA experiments. Under the Health and Safety at Work etc. Act, GMAG's advice had the force of law.

The United States equivalent is the RAC (Recombinant DNA Advisory Committee) which formulates the NIH Guidelines.

Goldberg–Hogness box *See* TATA box.

Goldbrick A slang term for a Goldberg–Hogness box.

Gram-negative One of the two categories into which all bacteria are divided according to their Gram's staining reaction. Bacteria are first stained with crystal violet and then with iodine. Treatment with acetone will destain Gram-negative bacteria. Such bacteria have a complex cell wall in which an outer membrane overlays the rigid peptidoglycan matrix. Enteric bacteria such as *Escherichia coli* and *Salmonella typhimurium* are Gram-negative.

The staining method is named after its inventor, Christian Gram, the nineteenth-century Danish bacteriologist.

Gram-positive Bacteria which resist decolorisation by acetone during the Gram's staining reaction (*see* Gram-negative). Gram-positive bacteria have a simpler cell wall than do Gram-negative bacteria. It is made predominantly of peptidoglycan and is not overlayed by an outer membrane. *Bacillus subtilis* and *Streptomyces* are genetically important Gram-positive bacteria.

Grunstein–Hogness procedure *See* colony hybridisation.

guide sequence An RNA molecule (or a part of one) which hybridises to eukaryotic mRNA and aids in the splicing of intron sequences. Guide sequences may be either external (EGS) or internal (IGS) to the mRNA being processed and may hybridise to either intron or exon sequences close to the splice junction.

Small nuclear RNA molecules, snRNAs or snurps, such as U1, act as external guide sequences (EGS) in mammalian cells. The rRNA precursor of *Tetrahymena* and pre-mRNAs in *Aspergillus* mitochondria are known to use internal guide sequences (IGS). (*See* ribozyme)

Fig. 21. Guide sequences.

External Guide Sequence (EGS)

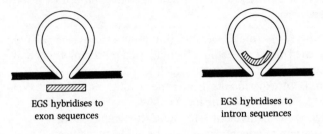

EGS hybridises to
exon sequences

EGS hybridises to
intron sequences

Internal Guide Sequence (IGS)

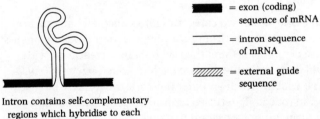

Intron contains self-complementary
regions which hybridise to each
other

= exon (coding)
sequence of mRNA

= intron sequence
of mRNA

= external guide
sequence

H

Hae **III** (pronounced hay three) A type II restriction enzyme from the
bacterium *Haemophilus aegyptius* which recognises the DNA sequence
shown below and cuts at the site indicated by the arrows:

$$\downarrow$$

5' G G C C 3'
3' C C G G 5'

$$\uparrow$$

(A full list of restriction enzymes can be found in Appendix 1)

hairpin loop A double-stranded structure formed by the base pairing of
self-complementary regions of a single DNA or RNA strand. E.g. the
sequence 5'—CGGTAATTACCG—3' can form a perfect hairpin loop
(Fig. 22).

Fig. 22. Hairpin loop.

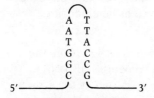

hairy root disease A disease of broadleaved plants in which there is a
proliferation of root-like tissue from the stem. This is a tumourous
state similar to the crown-gall. Hairy root disease is induced by the
bacterium *Agrobacterium rhizogenes* containing an Ri plasmid.

HART *See* hybrid arrested translation.

HAT medium A medium used in mammalian cell genetics to select
transformants or fusants which have an active thymidine kinase
enzyme (tk$^+$). The medium contains hypoxanthine, aminopterin and
thymidine. The two drugs block the conversion of dCDP to dTDP,
which is the normal route of thymidine biosynthesis in the cell. tk$^+$
cells will be able to grow in HAT medium by using their active
thymidine kinase enzyme to phosphorylate the thymidine supplied in
the medium. tk$^-$ cells will not be able to utilise this exogenous source
of thymidine and will die.

HCRM, host controlled restriction and modification The mechanism by
which certain bacteria attempt to protect themselves from phage

infection. Incoming phage (foreign) DNA is destroyed by site-specific restriction endonucleases. The bacterial genome is protected from destruction by its own restriction enzymes by the modification of some of the nucleotides in the enzyme's recognition site. Base methylation is the usual method of modification.

heavy metal resistance Resistance to the lethal effects of certain metals such as mercury, cadmium, lead, copper, zinc and tellurium. Resistance to mercury has been found on a transposon. Eukaryotes can exhibit heavy metal resistance via a class of proteins called metallothiens.

heteroduplex A DNA duplex prepared by the hybridisation of single-stranded DNA molecules derived from two different sources. Where the two DNAs have identical or very similar sequences, a double-stranded molecule will be established. However, where the two DNAs differ in sequence, single-stranded regions will remain. These will be revealed as single-strand 'bushes' when DNA is observed electron microscopically using the Kleinschmidt procedure. A map of homologous and non-homologous regions of the two molecules may thereby be constructed. This process is known as heteroduplex mapping.

Hha I (pronounced ha ha one) A type II restriction enzyme from the bacterium *Haemophilus haemolyticus* which recognises the DNA sequence shown below and cuts at the sites indicated by the arrows:

$$\downarrow$$
5′ G C G C 3′
3′ C G C G 5′
$$\uparrow$$

(A full list of restriction enzymes can be found in Appendix 1)

Hind III (pronounced hin dee three) A type II restriction enzyme from the bacterium *Haemophilus influenzae* Rd which recognises the DNA sequence shown below and cuts at the sites indicated by the arrows:

$$\downarrow$$
5′ A A G C T T 3′
3′ T T C G A A 5′
$$\uparrow$$

Hind III was one of the first type II restriction enzymes to be isolated and characterised and has been used for many purposes since. There is a single *Hind* III site in the tetracycline-resistance gene of pBR322. Many vectors have been constructed to have a *Hind* III cloning site. (A full list of restriction enzymes can be found in Appendix 1)

HnRNA, heterogeneous nuclear RNA High-molecular-weight RNA found in the nuclei of eukaryotic cells which is believed to represent the precursors to the mature mRNA molecules which are exported to the cytoplasm.

Hogness box *See* TATA box.

homocopolymer A nucleic acid chain containing only two sorts of nucleotide, usually in an alternating sequence, e.g. the RNA

$$5' \quad pApUpApUpApU \ldots \ldots pApU_{OH} \quad 3'$$

and the DNA duplex

$$5' \quad pGpT\,pGpT\,pGpT\,pGpT \ldots \ldots pGpT_{OH} \quad 3'$$
$$3' \quad _{OH}CpApCpApCpApCpAp \ldots \ldots CpAp \quad 5'$$

homogenotization A genetic technique where one copy of a gene, or other DNA sequence within a genome, is replaced with an altered copy of that sequence. The DNA is first cloned and then altered in some way, e.g. a transposon is inserted into a gene. The mutated gene copy can be used to replace the original gene by recombination *in vivo*. The incorporation of the mutated gene is usually selected, e.g. by virtue of its containing a transposon-encoded antibiotic resistance. (*See* replacement, transplacement)

homologous-assist *See* homologous helper plasmid.

homologous helper plasmid A plasmid used to assist in the uptake and establishment of a plasmid cloning vector in *Bacillus subtilis* and some other Gram-positive bacteria. The plasmid vector which is used to transform a *B. subtilis* strain will be linearised by the bacterium's DNA uptake apparatus and be destroyed unless it is rescued by recombination with a homologous plasmid within the cell. The homologous helper plasmid and the incoming plasmid each have a different antibiotic resistance gene so that selection for both resistances will ensure the survival of the desired DNA sequences. The plasmids are constructed to ensure that any foreign DNA carried by the incoming plasmid will be present in the final recombinant. This process is often known as plasmid rescue or homologous assist.

homology The degree of identity between the nucleotide sequences of two nucleic acid molecules or the amino acid sequences of two protein molecules. Although sequence determination is the ultimate test of homology, useful estimates can be provided by either DNA–DNA or DNA–RNA hybridisation.

homopolymer A nucleic acid chain containing only one sort of nucleotide, e.g. the RNAs, polyuridylic acid

5' pUpUpUpUpUpU . . . . pUpU$_{OH}$ 3'

and polyadenylic acid

5' pApApApApApA . . . . pApA$_{OH}$ 3'

are homopolymers, but the DNA duplex

5' pApApApApApA . . . . pApA$_{OH}$ 3'
3' $_{OH}$TpTpTpTpTpTp . . . . TpTp 5'

would also be described as a homopolymer.

homopolymer tail A stretch of homopolymer nucleic acid at the end of a DNA or RNA molecule, e.g. most eukaryotic mRNA has a homopolymer tail of pA at the 3' end.

host An organism able to support the replication of a plasmid or virus.

host-controlled restriction and modification *See* HCRM.

host range The number of different species or hosts in which a plasmid, phage or other virus will replicate.

Hpa I (pronounced heppa one or heapa one) A type II restriction enzyme from the bacterium *Haemophilus parainfluenzae* which recognises the DNA sequence shown below and cuts at the sites indicated by the arrows:

↓
5' G T T A A C 3'
3' C A A T T G 5'
↑

(A full list of restriction enzymes can be found in Appendix 1)

HRT *See* hybrid released translation.

hsdM The *E. coli* gene which encodes the DNA methylase involved in the modification system which prevents its DNA being attacked by its own restriction endonucleases. Host strains in gene cloning experiments usually have a functional *hsdM* gene to facilitate further gene transfers.

hsdR The *E. coli* gene which encodes restriction endonuclease activity. The initials stand for host specified defence restriction. Host strains for gene cloning experiments are often *hsdR*⁻ mutants so that the foreign DNA carried by the vector is not attacked by the host's restriction system. (*See hsdM*, HCRM)

hybrid arrested translation A method used to identify the proteins coded for by a cloned DNA sequence. A crude cellular mRNA preparation, composed of many individual types of mRNA, is hybridised with a cloned DNA. Only mRNA molecules homologous to the cloned DNA will anneal to it. The rest of the mRNA molecules are put into an *in vitro* translation system and the protein products are compared with the proteins obtained by use of the whole mRNA preparation.

hybridisation The formation of stable duplexes between complementary nucleotide sequences via Watson–Crick base-pairing. The efficiency of hybridisation is a test of sequence homology. DNA–DNA, DNA–RNA or, more rarely, RNA–RNA hybrids may be formed. (*See* Southern and Northern blotting)

An alternative use of the word comes from classical genetics, and particularly plant breeding. Here, hybridisation means to form a novel diploid organism either by normal sexual processes or by protoplast fusion.

hybridoma A hybrid cell-line produced by the fusion of a normal lymphocyte with a myeloma cell. (Myelomas are tumours of the immune system in which a single lymphocyte line proliferates in an uncontrolled manner.) Following selection and cloning an individual hybridoma line will produce only one type of antibody, a monoclonal antibody.

hybrid released translation A method used to detect the proteins coded for by cloned DNA. The cloned DNA is bound to a nitrocellulose filter and a crude preparation of mRNA is hybridised to the filter-bound DNA. Only mRNA sequences homologous to the cloned DNA will be retained on the filter. These mRNA molecules can then be removed by high temperature or by using formamide. The purified mRNA is then placed in an *in vitro* translation system and the proteins encoded by the message can be analysed by electrophoresis through a polyacrylamide gel.

hydrogen-bonding A non-covalent bond formed between the hydrogen atom in an —O—H or N—H group and an oxygen or nitrogen atom. Such bonds are essentially ionic in nature, the hydrogen atom bearing a partial positive charge, due to the withdrawal of electrons by its neighbouring N or O atom, and the oxygen or nitrogen atom bearing a partial negative charge. Hydrogen bonds are weak but are very important in the maintenance of the secondary structure of both proteins and nucleic acids. It is hydrogen bonds which maintain the double helical structure of DNA and ensure the fidelity of DNA replication, transcription and mRNA translation.

The hydrogen bonds involved in DNA base pairing are shown below. GC base pairs involve three hydrogen bonds while AT base pairs have only two. This means that DNA molecules with a high proportion of

GC will be more stable and not so readily denatured as those rich in AT.

Fig. 23. Hydrogen bonding (structure of base pairs).

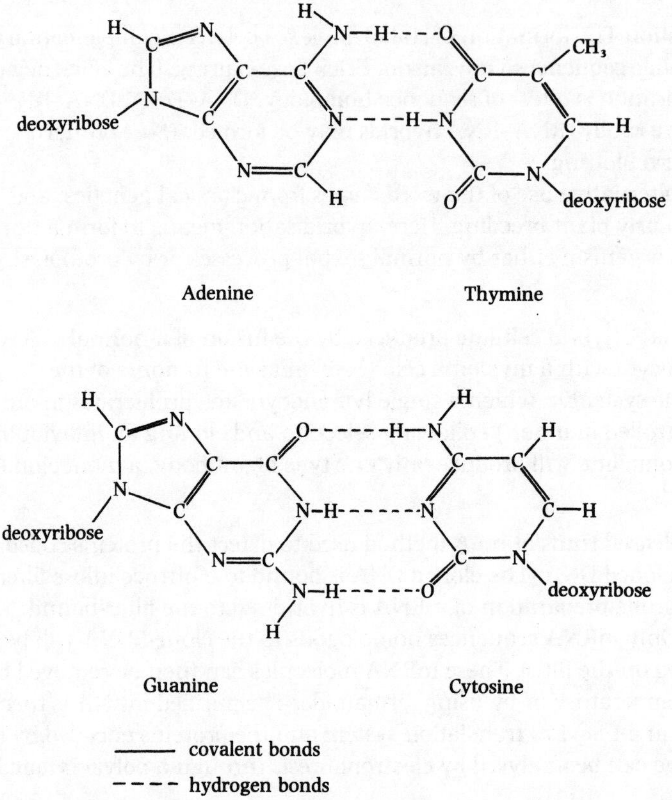

Adenine   Thymine

Guanine   Cytosine

——— covalent bonds
– – – – hydrogen bonds

hydroxylapatite, hydroxyapatite, HAP A form of calcium phosphate which can bind nucleic acids. Under certain conditions it will bind double-stranded but not single-stranded DNA. It can therefore be used to fractionate DNA preparations and to determine the extent of hybridisation of two classes of single-stranded molecules.

I

IGS *See* guide sequence.

illegitimate recombination Recombination between DNA species which show very little or no homology. Illegitimate recombination is promoted by transposons and IS elements.

immunity (i) In bacteriophage genetics immunity is the state where a bacterium which carries prophage in the host chromosome will not support the replication of a superinfecting phage of the same type. The bacterium carrying the prophage (lysogen) is thus immune. The prophage directs the synthesis of repressor proteins which recognise operators on the superinfecting phage and thus repress transcription. (ii) In transposition a replicon which carries one copy of a transposon will not accept another copy of that transposon by transposition. This applies to several, but not all, transposons. (iii) The resistance of an organism producing an antibiotic (such as a bacteriocin) to the lethal effects of that antibiotic. Col-factors are plasmids which contain genes determining both colicin production and immunity.

immunoassay An assay system which detects proteins by using an antibody specific to that protein. A positive result is seen as a precipitate of an antibody–protein complex. The antibody can be linked to a radioactive atom (*see* radioimmunoassay) or to an enzyme which catalyses an easily monitored reaction (*see* ELISA).

Inc *See* incompatibility.

incompatibility (i) A function of a related group of plasmids. Plasmids which are closely related share similar replication functions and this leads to the exclusion of one or the other plasmid if they are present in the same cell; thus such plasmids are incompatible. Plasmids are placed in incompatibility groups by this simple reaction and, in general, plasmids belonging to one incompatibility group are very closely related. Such groups are called Inc A, Inc B, Inc C, etc. (ii) In fungi which have a sexual cycle two different mating types must fuse to establish the heterokaryon or diploid stage. If the two mating types are the same, incompatibility exists and fusion cannot occur or is abortive.

incomplete digest *See* partial digest.

inducer A chemical or physical agent which, when given to a population of cells, will increase the amount of transcription from specific genes. E.g. isopropylthio-β-galactoside is a powerful inducer of the *lac* operon.

inducible A gene or gene-product is said to be inducible if its transcription or synthesis is increased by exposure of the cells to an effector. These are usually small molecules whose effects are specific to particular operons or groups of genes (*cf.* constitutive). (*See lac*)

initiation codon, initiator The codon which specifies the first amino acid of a polypeptide chain. In bacteria, the initiation codon is either AUG, which is translated as *n*-formyl methionine (a modified amino acid) or, rarely, GUG (valine). In eukaryotes the initiation codon is always AUG and is translated as methionine. The term is also used to describe the corresponding sequence in DNA, ATG.

insert The piece of foreign DNA introduced into a vector molecule.

insertional inactivation The technique in which foreign DNA is cloned into a restriction site which lies within the coding sequence of a gene in the vector. The insertion of foreign DNA at such a site interrupts the gene's sequence such that its original function is no longer expressed. This permits the detection of recombinant molecules following transformation. E.g. cloning into the *Pst* I site of the vector pBR322 inactivates the ampicillin-resistance gene but leaves the tetracycline-resistance gene intact. Following transformation, cells carrying recombinant molecules will have an ampicillin-sensitive, tetracycline-resistant phenotype. (Fig. 24)

insertion sequence A class of small DNA elements which can transpose. In general, they contain no phenotypic marker and were first identified due to their ability to insert into a gene and destroy its function. Insertion sequences were the first transposons to be identified in bacteria. Their coding capacity is sufficient only for one or two proteins. Some insertion sequences contain a promoter near their termini which can allow transcription of DNA next to the insertion sequence. Insertion sequences have been named IS1, IS2, etc.

insertion site, cloning site A unique restriction site in a vector DNA molecule into which foreign DNA can be inserted. The term is also used to describe the position of integration of a transposon or IS element.

insertion vector A phage-cloning vector, usually λ, which has unique restriction sites in a non-essential region. λ phages need to be between 75 and 105% of the size of wild-type λ DNA to be packaged and

Fig. 24. Insertional inactivation.

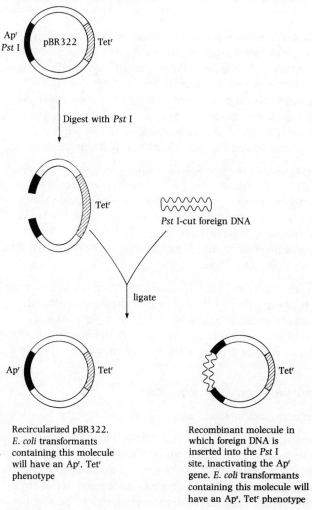

Recircularized pBR322. *E. coli* transformants containing this molecule will have an Ap^r, Tet^r phenotype

Recombinant molecule in which foreign DNA is inserted into the *Pst* I site, inactivating the Ap^r gene. *E. coli* transformants containing this molecule will have an Ap^s, Tet^r phenotype

insertion vectors are at the lower end of these size limits. These vectors therefore can accommodate *ca.* 20% of the size of wild-type λ, i.e. up to 10 kb of insert DNA. The site into which foreign DNA is cloned is often within a gene which enables recombinant phage plaques to be distinguished from plaques formed by phages which contain no insert DNA.

integration The recombination process which inserts a small DNA molecule into a larger one. If the molecules are circular this involves only a single cross-over, if linear then two cross-overs are required. A

55

well known example is the integration of phage λ DNA into the *E. coli* genome.

intensifying screen A plastic sheet impregnated with a rare-earth compound, e.g. calcium tungstate. The rare-earth compound will absorb β radiation and emit light. When placed on one side of a piece of X-ray film with a radioactive sample on the other side, the intensifying screen will capture some of the β emissions which pass through the film. The light emitted by the screen upon β-capture will blacken the X-ray film and so greatly enhance the sensitivity of the detection. It is used in Southern and Northern blotting experiments.

intercalate Certain drugs or dyes, such as ethidium bromide, are able to insert into DNA or double-stranded RNA between adjacent base pairs. Molecules with this property are called intercalating agents. The binding of such dyes reduces the buoyant density of the DNA. The DNA duplex increases in length and, if the DNA is supercoiled, increasing concentrations of the dye first unwind the supercoils and then wind the molecule up again in the opposite sense.

intercalating agent These are planar molecules which are able to insert themselves between adjacent base pairs in a double-stranded DNA or RNA molecule. Intercalating agents inhibit the replication and transcription of DNA, promote the curing of plasmids from their hosts and reduce the buoyant density of DNA in solution. (*See* ethidium bromide)

internal guide sequence *See* guide sequence.

intervening sequence *See* intron.

intron, intervening sequence A sequence within a eukaryotic gene which is not represented in the protein product of that gene. Intron sequences are transcribed into RNA and must be excised, and the RNA molecule religated, before it can be translated; a process known as intron splicing. Some genes of higher eukaryotes contain a large number of introns which make up the bulk of the DNA sequence of the gene. Introns are also found in genes whose RNA transcripts are not translated, i.e. eukaryotic rRNA and tRNA genes. In these cases the intron sequence does not appear in the functional RNA molecule (*see* exon).

inverted repeat Two regions of a nucleic acid molecule which have the same nucleotide sequence but in the opposite orientation to one another (*cf.* direct repeat). E.g.

GCACTTG. . . . . GTTCACG
CGTGAAC. . . . . CAAGTGC

in vitro **packaging** The formation of a functional viral particle around phage λ or cosmid DNA *in vitro*. A preparation of semi-complete, preformed phage head and tail components are made from two mutant λ lysogens. When these are mixed *in vitro* with any concatemeric DNA which contains two *cos* sites separated by about 49 kb of DNA, that DNA will be packaged into an infective viral particle. The infection of *E. coli* with such particles represents a highly efficient method of introducing recombinant DNA, in either cosmid or λ cloning vectors, into the organism.

in vitro **transcription, cell-free transcription** The specific and accurate synthesis of RNA in the test-tube using purified DNA preparations as a template. So-called 'coupled systems' may be obtained from *E. coli* which carry out both mRNA synthesis and its translation into protein. For eukaryotes, separate cell-free systems have to be set up to demonstrate the activity of the three functionally distinct RNA polymerase complexes.

in vitro **translation, cell-free translation** The synthesis of proteins in the test-tube from purified mRNA molecules using cell extracts containing ribosomal subunits, the necessary protein factors, tRNA molecules and aminoacyl tRNA synthetases. ATP, GTP, amino acids and an enzyme system for regenerating the nucleoside triphosphates are added to the mix. Prokaryotic translation systems are usually prepared from *E. coli* or the thermophilic bacterium *Bacillus stearothermophilus*. Eukaryotic systems usually employ rabbit reticulocyte lysates or wheatgerm.

IS element *See* insertion sequence.

isopycnic centrifugation *See* density gradient centrifugation.

isoschizomer (pronounced isoskitzomer) Two restriction enzymes which have the same target sequence are described as a pair of isoschizomers. E.g. *Hpa* II and *Msp* I both cut at

```
         ↓
5′   C  C  G  G   3′
3′   G  G  C  C   5′
         ↑
```

i-value The total concentration of DNA termini in a ligation reaction. This value increases with increasing DNA concentration. *i*-values and *j*-values influence the formation of products in a ligation reaction. By choosing different *i*- and *j*-values, i.e. various concentrations of vector and insert DNAs, one can influence the structures of the recombinant molecules formed. These may be linear oligomers, circularized vector or vector plus insert. (*See j*-value for a worked example)

J

j-value The effective concentration of one end of a DNA molecule in the immediate neighbourhood of the other end of the same molecule. j-values and i-values (the total concentration of ends in the reaction mix) influence the formation of products in a ligation reaction. The j-value is a function of the length of a DNA molecule and is independent of the concentration of DNA. When $j/i \leqslant 1$, linear oligomers are formed. When $j/i \geqslant 2$, circular molecules are formed. When j/i is between 1 and 1.8, linear oligomers are formed initially which subsequently circularize as the j/i value rises during the course of the reaction.

Example for linearised pBR322:

j for pBR322 $= 12.5 \times 10^{-12}$ ends ml^{-1}

i = total concentration of ends in the reaction mixture

pBR322 at 100 μg ml^{-1} $i = 70.6 \times 10^{-12}$ ends ml^{-1} $j/i = 0.177$

100 μg ml^{-1} $i = 7.06 \times 10^{-12}$ ends ml^{-1} $j/i = 1.77$

1 μg ml^{-1} $i = 0.706 \times 10^{-12}$ ends ml^{-1} $j/i = 17.7$

j/i 0.177, linear oligomers are formed

j/i 1.77, linear oligomers which subsequently circularize

j/i 17.7, circularized monomers are formed.

K

kanamycin resistance, Km^r or Kan^r; kanamycin sensitive, Km^s or Kan^s
Resistance or sensitivity to the lethal effects of the aminoglycoside antibiotic, kanamycin. Some cloning vectors have a kanamycin-resistance gene as a selectable marker. One such gene also determines resistance to the related antibiotics neomycin and G418. The latter does not inhibit the growth of bacteria but is active against eukaryotes. For this reason the gene encoding G418 resistance has been incorporated into some eukaryotic cloning vectors.

kb An abbreviation for kilobase pair.

kilobase pair One thousand base pairs of DNA duplex. A convenient measure of DNA molecular weight.

kinase An enzyme which will remove or add a phosphate group to a protein or nucleic acid. (*See* T4 polynucleotide kinase)

Kleinschmidt spread A technique by which minute quantities of nucleic acids are coated with a basic protein and spread on a denatured protein monolayer at an air–water interface. They are then shadowed with a heavy metal prior to being viewed in an electron microscope. The technique therefore permits the visualisation of single and double-stranded RNA or DNA molecules in the electron microscope. Without the layers of protein and heavy metal atoms, these molecules are too thin to be resolved by the electron microscope. The thickness of the coated molecules is proportional to their original diameter and thus single- and double-stranded molecules can be distinguished.

Klenow fragment, Klenow enzyme (pronounced Klenof) The larger of the two fragments of *E. coli* DNA polymerase I formed after cleavage with a protease such as subtilisin. The large fragment retains the 5′ → 3′ polymerase and the 3′ → 5′ exonuclease activities and can therefore be used in the Sanger method of DNA sequencing. It cannot be used for nick translation since it does not have the 5′ → 3′ exonuclease activity required to extend a nick.

Kpn I A type II restriction enzyme from the bacterium *Klebsiella pneumoniae* OK8 which recognises the DNA sequence shown below and cuts it at the sites indicated by the arrows:

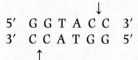

(A full list of restriction enzymes can be found in Appendix 1)

L

lac The three letter symbol for the genotype concerned with lactose utilisation and/or uptake. In *E. coli* the *lac* genes are organised into an operon of *lacZYA* which encode respectively, the β-galactosidase, lactose permease and lactose transacetylase enzymes. (*See* fusion protein, M13mp vectors)

lacZ **fusion** *See* fusion protein.

λ, **phage lambda** The best-studied of all the temperate bacteriophages and the basis for many of the most popular phage vectors for gene cloning. A mature λ particle consists of an icosohedral protein head and a long tail. The phage head contains a linear double-stranded DNA molecule of 49 kb in length. The two 5′ ends of the molecule form 12 bp sticky ends which permit the circularization of the phage DNA when injected into a new host.

In the lytic cycle of infection, many daughter genomes are synthesised by the rolling-circle mechanism of replication. The multimer produced by this mode of replication is cut into its constituent monomers by the *ter* (for termination) protein. This protein is analogous to a type II restriction endonuclease and recognises the nucleotide sequence of the *cos* (for cohesive) site and cleaves within it to generate the 12 bp sticky ends. The monomers produced must have a size of *ca.* 44–52 kb to be successfully packaged into an infective phage particle.

In the lysogenic cycle, the replication of the infecting λ DNA is repressed and instead, the circular molecule recombines with the *E. coli* chromosome at a site between the *gal* (galactose) and *bio* (biotin) genes. The phage genome is therefore inserted into the host chromosome and behaves like any other genetic locus within that chromosome, as long as repression is maintained.

λ **exonuclease** An exonuclease encoded by bacteriophage λ which will remove DNA from the 5′ end of a duplex. It is sometimes used to provide a good substrate for terminal transferase.

λ**gt.λC** A phage λ replacement vector. The 'gt' stands for generalised transducer and signifies that this vector will act as a generalised transducing phage for any foreign DNA inserted into the central replaceable region.

The central, non-essential portion of this vector has the *Eco* RI C fragment from the parent phage, hence the 'λC'.

The central dispensable C fragment can be replaced by any foreign DNA fragment of equivalent size. Phage λ molecules have to be between 75 and 105% of the size of wild-type λ DNA to be correctly packaged and this means that there is a positive selection for the insertion of any piece of DNA of the correct size between the two *Eco* RI sites in the centre of this vector. The left arm and the right arm alone are too small to be packaged.

λgt.WES A series of phage λ replacement vectors based on the λgt.λC vector and equivalent to λWES.λB' vector. The *Eco* RI C fragment of phage λ carries genes for specialised recombination, and replacement of this fragment by the *Eco* RI B fragment led to a vector with a higher degree of biological containment. (*See* λWES)

λplac (pronounced pee lac) A λ bacteriophage which carries a segment of the *E. coli lac* operon. The *p* stands for the ability to form plaques.

λWES A λ phage cloning vector which has amber mutations in the W, E and S genes. These mutations afford a high degree of biological containment since this phage can only be propagated in a special host.

λWES·λB' A phage λ replacement vector which carries a dispensable *Eco* RI fragment, *Eco* RI B from the parent phage. The designation B' indicates that the orientation of the B fragment is reversed when compared to the parent phage. Since the right and left arms of the vector carry all the essential genes for plaque formation the central *Eco* RI B' fragment can be replaced by foreign DNA of equivalent size. λ phages have to be between 75 and 105% of the size of wild-type λDNA to be packaged, hence there is a positive selection for the presence of a piece of foreign DNA in the centre of this vector.

lariot *See* banjo.

leader sequence, leader peptide *See* signal sequence.

lethal zygosis A term borrowed from classical genetics used to describe the formation of pocks in a lawn of Actinomycetes. Certain plasmids of *Streptomyces* species contain a gene or genes responsible for pock formation and these are known as *ltz* for lethal zygosis.

LGT agarose *See* low-gelling-temperature agarose.

library, gene library *See* bank.

ligase An enzyme which seals nicks in DNA molecules by forming a phosphodiester bond between adjacent nucleotides which have free 5'-phosphate and 3'-hydroxyl groups. The ligase enzyme encoded by phage T4 is usually used in gene-cloning experiments; it requires ATP as a cofactor. T4 ligase is used *in vitro* to join vector and insert DNA. (*See* RNA ligase)

ligation The process of joining two linear nucleic acid molecules together via a phosphodiester bond. In a cloning experiment a restriction fragment is often ligated to a linearised vector molecule using T4 DNA ligase.

linear A term used to describe the physical state of a nucleic acid molecule. The two strands at the end of a linear molecule of double-stranded DNA can be free, as in a restriction fragment, bound to a specific protein, as in some eukaryotic viruses, or closed by a hairpin loop, as in some eukaryotic extrachromosomes. Eukaryotic chromosomes are linear and their ends, the telomeres, have special sequences which promote their replication.

linearise To introduce a single double-strand break in a covalently closed circular DNA molecule and so convert it to a linear molecule. The cleavage of a unique restriction site within a plasmid will linearise it.

linker A synthetic oligodeoxyribonucleotide which contains a restriction site. Linkers may be blunt end-ligated onto the ends of DNA fragments to create restriction sites which can be used in the subsequent cloning of the fragment into a vector molecule. It may be necessary to protect the fragment to be cloned from the action of the restriction endonucleases by treating them with DNA methylase. An example of a synthetic linker is the decamer:

5′   CCGAATTCGG
3′   GGCTTAAGCC

This contains an *Eco*R1 site (boxed). Self-ligation of linker molecules to form multimers generates restriction sites for the blunt-end cutter *Hae*III (boxed below) and thus permits the regeneration of linker monomers:

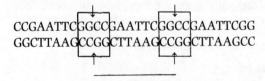

CCGAATTCGGCCGAATTCGGCCGAATTCGG
GGCTTAAGCCGGCTTAAGCCGGCTTAAGCC

linker monomer

liposome An artificial membrane vesicle consisting of a spherical phospholipid bilayer. DNA molecules may be entrapped in, or bound to the surface of, the vesicles and subsequent fusion of the liposome with the cell membrane will deliver the DNA into the cell. Liposomes have been used to develop an efficient transfection procedure for *Streptomyces* bacteria.

localised random mutagenesis *See* site-specific mutagenesis.

locus Any site which has been defined genetically. A locus may be a gene, a part of a gene, or a DNA sequence which has some regulatory role. (*See* marker)

low-gelling-temperature agarose, LGT-agarose, low-melting-point agarose An agarose derivative which will remain fluid at temperatures around 37 °C. It also melts at low temperature, *ca.* 60–65 °C and this means that DNA fragments can readily be recovered in their native form following electrophoresis.

lysogenic phage A bacteriophage which, instead of always lysing its host, can sometimes enter into a stable relationship with it, its genes being replicated in phase with the bacterial chromosome. To achieve this, the phage DNA is either integrated into the host chromosome or exists as a plasmid. A bacterium harbouring a phage genome in this state is said to be a lysogen.

lysogeny The developmental pathway of a temperate bacteriophage in which lytic functions are repressed and the phage DNA replicates coordinately with that of the host chromosome. In lysogeny the phage DNA may either be integrated into the host chromosome, e.g. λ, or remain as an independent episome, e.g. P22.

lysozyme An enzyme which degrades the rigid cell wall material (peptidoglycan) of many bacteria. Bacteria treated with this enzyme can be lysed with detergents.

lytic infection The developmental pathway in which bacteriophage DNA enters a host bacterium and is replicated in an uncontrolled manner. The host chromosome is degraded to provide nucleotides for phage DNA replication and the host's protein-synthetic machinery is taken over to make the components of the mature phage particle. When a large number of phage genomes have been packaged into particles the host cell is lysed to release mature phage which may go on to infect other bacteria.

M

map As a verb: to determine the relative positions of genes or restriction sites on a DNA molecule. Genetic mapping is done by carrying out mating experiments to determine the linkage relationships between genes and the frequency of recombination between them. The further apart two genes are on a chromosome, the greater will be the frequency of recombination between them; if they are on different chromosomes, i.e. they are unlinked, they will show the maximum recombination frequency of 50%. Physical mapping is usually performed by carrying out a number of single and double digests with a range of restriction enzymes.

As a noun: a diagram showing the relative positions of, and distances between, genes or restriction sites.

marker A mutation in a gene which facilitates the study of its inheritance.

Maxam and Gilbert method, chemical method of DNA sequencing A method of DNA sequence analysis which uses chemical reactions to partially cleave a DNA fragment. The fragment, labelled at one end only with ^{32}P, is divided into four aliquots. Each aliquot is partially degraded using chemical reactions which are specific for one type of nucleotide. When these samples are separated by electrophoresis through a polyacrylamide gel they form a nested set of lengths one end of which is always the end labelled with ^{32}P, the other end terminates at the particular nucleotide to which the chemical reaction was specific. These labelled fragments can be detected by autoradiography and the DNA sequence can be read directly from the band pattern on the X-ray film. (Fig. 25)

maxi-cell An *E. coli* or *B. subtilis* cell in which plasmid genes are preferentially expressed because the chromosomal genes have been inactivated by ultraviolet (UV) irradiation. A bacterial strain carrying mutations which render it defective in both DNA repair and recombination is exposed to UV light. Because of these deficiencies the chromosomal DNA will be severely damaged and degraded. If such a cell contains a multicopy plasmid-cloning vector some of the plasmids will not be hit by UV and these will continue to replicate and any proteins which they encode will continue to be produced. A radioactively labelled amino acid can be added to the irradiated culture and will be specifically incorporated into plasmid-encoded proteins.

The proteins encoded by λ cloning vectors can be studied by first irradiating the host cells with UV and then infecting them with the λ cloning vector.

Fig. 25. Maxam and Gilbert method.

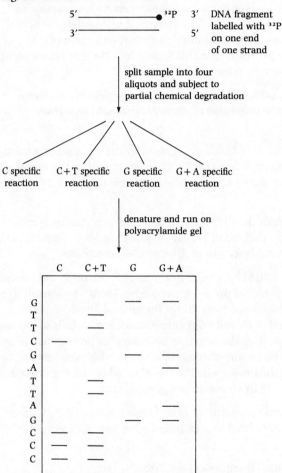

Md, megadalton 10^6 daltons of molecular weight; MW (or Mr) 10^6. A convenient shorthand used in expressing the molecular weight of DNA molecules.

melting temperature, T_m The temperature at which a double-stranded DNA or RNA molecule denatures into separate single strands. The T_m is characteristic of each DNA species and gives an indication of its base composition. This is because DNAs rich in G·C base pairs are more resistant to thermal denaturation than A · T rich DNA since three hydrogen bonds are formed between G and C, but only two between A and T.

methotrexate^r Resistance to the drug methotrexate (amethopterin) is a popular selectable marker in cloning experiments with mammalian cells. The drug is an inhibitor of the enzyme dihydrofolate reductase (DHFR) and one route to resistance is the overproduction of this enzyme. Thus the inclusion of a DHFR gene in a cloning vector permits the selection of methotrexate^r transformants.

methylase A DNA modification enzyme which covalently attaches a methyl (CH_3-) group to specific nucleotide bases within a DNA molecule.

methylation The process of base modification by the addition of a methyl (CH_3-) group. Methylation of specific nucleotides within the target site of a restriction enzyme can protect the DNA against attack by that enzyme.

methylmercuric hydroxide, CH_3HgOH A powerful denaturing agent which may be incorporated into agarose gels in order to determine the single-strand molecular weight of RNA or DNA molecules.

mini-cell These are formed by special mutants of *E. coli* or *B. subtilis* and are small cells which contain no chromosomal DNA. They result from a cell wall being laid down near one of the ends of a rod-shaped bacterium. Plasmid DNA will segregate into such mini-cells along with ribosomes, tRNA and all the enzymes necessary for protein synthesis. Mini-cells cannot make any chromosomally encoded proteins and are used for studying the proteins defined by plasmid genes, e.g. from a segment of foreign DNA cloned into a plasmid vector.

mini-prep A small-scale preparation of plasmid or phage DNA. Usually used to analyse insert DNA in a cloning vector after a cloning experiment.

−10 sequence (minus 10 sequence) *See* Pribnow box.

−35 sequence (minus 35 sequence) A region of DNA upstream of prokaryotic promoters which is centred about 35 nucleotides from the mRNA-initiation site. Most prokaryotic promoters show some conserved sequences in the −35 region. The sequence

5′	T	T	G	A	C	A	3′
3′	A	A	C	T	G	T	5′
	−36	−35	−34	−33	−32	−31

is highly conserved. The -35 sequence is thought to be involved in the initial recognition between RNA polymerase and the promoter site.

mob The genetic notation for gene(s) involved in the mobilisation of a plasmid from cell to cell. Certain broad-host-range cloning vectors are mob^+ to enable them to be mobilised into a large variety of Gram-negative bacteria.

mobilisation (i) The transfer, between bacteria, of a non-conjugative plasmid by a conjugative plasmid. (ii) The transfer, between bacteria, of chromosomal genes by a conjugative plasmid.

modification Any post-synthetic chemical change to a DNA molecule. Common modifications are methylation and glucosylation – the covalent attachment of methyl and glucose moieties. In host-controlled restriction and modification (HCRM), the DNA of the host is protected against the action of its own restriction enzymes by such modifications.

monoclonal antibody An antibody preparation which contains only a single type of antibody molecule. Monoclonal antibodies are produced naturally by myeloma cells. A myeloma is a tumour of the immune system. A clone of cells producing any single antibody type may be prepared by fusing normal lymphocyte cells with myeloma cells to produce a hybridoma. The technique is outlined in Fig. 26.

mRNA, messenger RNA The RNA transcript of a protein-encoding gene. The information encoded in the mRNA molecule is translated into a polypeptide of specific amino acid sequence by the ribosomes. In eukaryotes, mRNAs transfer genetic information from the genes, in the nucleus, to the ribosomes, in the cytoplasm.

M13 A phage which attacks *E. coli* cells harbouring the F sex factor. The M13 does not lyse the host cells but virus is continuously extruded into the medium. The virus slows cell growth enough for the infected cells to be recognised as plaques. The viral nucleic acid is a single-stranded circle of DNA. The intracellular replicative form (RF) is a double-stranded circle of DNA which can be isolated in the same way as a plasmid. Several derivatives of M13 have been constructed *in vitro* to become a cloning/sequencing system. These derivatives, called mp2–mp11, all contain a portion of the *E. coli lac* operon and a selection of single restriction enzyme sites for cloning of foreign DNA. The portion of the *lac* operon gives a visual indication of the presence of an insert since the cloned fragment interrupts the α-peptide portion of the β-galactosidase gene and changes the colour of a phage plaque from blue to white when an indicator (X-gal) is incorporated into the medium. Large quantities of single-stranded DNA can be isolated from virus in the medium; this gives a pure template for DNA sequencing by the Sanger method.

Fig. 26. Monoclonal antibody.

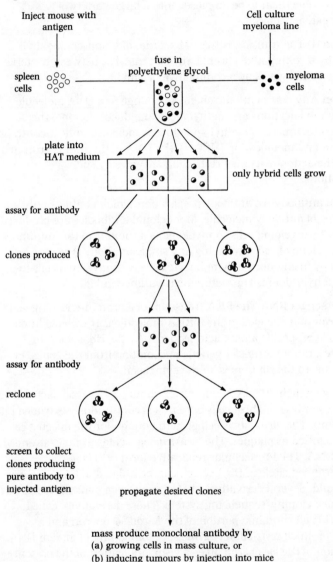

Mu, µ A DNA-containing temperate bacteriophage which can integrate into the host cell's DNA at a great number of sites and lead to a mutation if it integrates into a gene, hence Mu from 'mutator phage'. It is thought to replicate by a process of transposition and may therefore be considered as a transposon.

µ, µm, micron A unit of length, equal to 10^{-6} metre, which is often used to describe DNA molecules. 1 µ of DNA duplex is *ca.* 2×10^6 daltons which is *ca.* 3 kb.

multi-copy An adjective used to describe plasmids which replicate to produce many plasmid molecules per host genome, e.g. pBR322 is a multi-copy plasmid, there are usually 50 pBR322 molecules (or copies) per *E. coli* genome.

multimer A complex made up of a number of, sometimes identical, protein or nucleic acid molecules.

N

NACS™ (pronounced nacks to rhyme with sacks) An ion-exchange resin to which nucleic acids can be quantitatively bound in low-salt buffers and then released by washing with a high-salt buffer. NACS columns are used to free DNA or RNA from contaminating low-molecular-weight compounds and are especially useful for purifying DNA which has been extracted from polyacrylamide or agarose gels. (NACS™ is a trademark of Bethesda Research Laboratories)

negative control (i) In genetics, a type of control on the transcription of a gene or operon where a repressor protein binds to an operator site upstream from the coding region and prevents transcription by RNA polymerase. In many cases the repressor can also bind another molecule, the inducer, which causes the repressor to lose its affinity for the operator sequence, thus allowing transcription to take place. (ii) In enzymology, the activity of an enzyme under negative control is reduced when it binds to an effector molecule.

nick A break or cut in one of the two strands of a DNA duplex.

nick translation A technique for radioactively labelling a DNA molecule to high specific activity in order to produce a probe which may be used in some hybridisation technique such as Southern blotting. Nicks or gaps are introduced into the DNA molecule using a limited DNase digest. Each nick is then extended by the 5′ to 3′ exonuclease activity of DNA polymerase I. At the same time as 5′ nucleotides are being removed by the exonucleolytic activity of the enzyme, they are also replaced by its polymerising activity. The supply of α-labelled ^{32}P deoxynucleotide triphosphates to the reaction ensures that the newly synthesised DNA will be highly radioactive. (*See* DNA polymerase I)

nif The genetic notation for the genes involved in nitrogen fixation. The *nif* operon is a complex array of seventeen genes. The proteins encoded by the *nif* genes will fix atmospheric nitrogen (N_2) into ammonia (NH_4^+) and nitrate (NO_3^-). Many soil bacteria will fix nitrogen and there is much interest in manipulation of *nif* genes from bacteria to allow plants to fix nitrogen.

NIH, National Institutes of Health A group of research institutes located in Bethesda, Maryland and also the principal US Government agency for funding biomedical research in universities.

NIH Guidelines Recommended procedures for the conduct of recombinant DNA experiments to which all NIH grant-holders must adhere.

nitrocellulose, cellulose nitrate A nitrated derivative of cellulose which is made into membrane filters of defined porosity, e.g. 0.45 μm, 0.22 μm. These filters have a variety of uses in molecular biology, particularly in nucleic acid hybridisation experiments. In the Southern and Northern blotting procedures DNA and RNA, respectively, are transferred from an agarose gel to a nitrocellulose filter.

Some centrifuge tubes are made of nitrocellulose; they are readily punctured with a hypodermic needle, and are frequently used in sucrose gradient centrifugation. (*See* Biodyne, colony hybridisation)

nonsense mutation A mutation which converts a codon which specifies an amino acid into a stop codon, e.g. a change from UAU (tyr) to UAG (amber) would lead to the premature termination of a polypeptide chain at the place where a tyrosine was inserted in the wild-type. (*See* stop codon, suppressor)

nopaline A rare amino acid derivative which is produced by a certain type of crown-gall tissue. The genes responsible for the synthesis of nopaline are part of the T-DNA from a Ti-plasmid.

Northern blot, Northern transfer A procedure analogous to Southern transfer but in this case RNA, not DNA, is transferred or blotted from a gel to a suitable binding matrix, such as a nitrocellulose sheet. Single-stranded RNA is separated according to size by electrophoresis through an agarose or polyacrylamide gel, the RNA is then blotted directly on to the support matrix with no denaturation. RNA fixed to the supporting matrix can then be hybridised with a radioactive single-stranded DNA or RNA probe.

nuclease Any enzyme which will hydrolyse a phosphodiester bond in a nucleic acid molecule. Nucleases are usually specific for DNA or RNA and for either single-stranded or double-stranded molecules. Some release nucleotides from the end of the molecule (exonucleases), others cleave the polynucleotide chain at an internal site (endonucleases). Nucleases have varying degrees of base sequence specificity, the most specific being the restriction endonucleases. (*See* BAL-31, complete digest, endonuclease, exonuclease, exonuclease III, four base pair cutter, restriction enzyme, RNase, six base pair cutter, S1 nuclease, target site)

nucleic acid A DNA or RNA molecule which can be single-stranded or double-stranded. When very small it is called an oligonucleotide.

nucleoside, deoxyribonucleoside, ribonucleoside A purine or pyrimidine base covalently bound to either a deoxyribose or a ribose sugar moiety, but without any phosphate groups.

nucleosome Eukaryotic chromatin has been found to have a 'beads-on-a-string' structure with regions of DNA tightly complexed with protein, separated by relatively naked stretches of the nucleic acid. The 'beads' are termed nucleosomes and consist of *ca.* 145 bp of DNA wound around a core particle which is an octomer of two molecules each of histones H2A, H2B, H3 and H4.

nucleosome phasing The arrangement of the nucleosomes with respect to specific DNA sequences within a eukaryotic chromosome. Nucleosome phasing is thought to control the expression of some eukaryotic genes. If the DNA sequence containing a gene's promoter is buried within a nucleosome it is inaccessible to RNA polymerase and the gene is not expressed. On the other hand, if the gene's promoter is contained in an internucleosome space, then RNA polymerase may bind to the promoter and express the gene. Changes in nucleosome phasing have been correlated with developmental switches in some eukaryotic organisms.

nucleotide, deoxyribonucleotide, ribonucleotide A nucleoside with one or more phosphate groups esterified to the 5′ position of the sugar moiety.

O

octopine A rare amino acid derivative which is produced by a certain type of crown-gall tissue. The genes responsible for the synthesis of octopine are part of the T-DNA from a Ti-plasmid.

Okazaki fragment DNA polymerase enzymes are able to synthesise DNA only in the 5′ to 3′ direction. As the replication fork moves along the parental duplex, one daughter strand can be synthesised continuously but the other strand must be synthesised in short pieces as the polymerase complex jumps forward and synthesises DNA in the opposite direction to the one in which the fork is moving. These short pieces are 1–2 kb in length and are called Okazaki fragments after the Japanese scientist who discovered them (Fig. 27).

Fig. 27. Okazaki fragment.

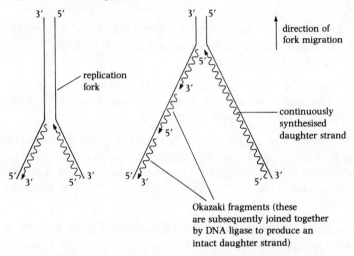

Okazaki fragments (these are subsequently joined together by DNA ligase to produce an intact daughter strand)

oligo(dT) cellulose Short lengths of deoxythymidilic acid(dT) homopolymer covalently attached to powdered cellulose. Oligo(dT) cellulose is used in the purification of eukaryotic mRNA by affinity chromatography. In this process the poly(A)-tail of the mRNA molecule is immobilised by hybridisation with the oligo(dT).

oligonucleotide A short-chain nucleic acid molecule.

oligonucleotide-directed mutagenesis *See* site-specific mutagenesis.

o-micron DNA An alternative name for the 2 μ plasmid of *Saccharomyces cerevisiae*.

73

open circle One of the three conformations which a plasmid molecule may adopt. In an open circle there has been either a single break or a number of staggered breaks made in the strands of the duplex. Such a molecule may not form supercoils and therefore is referred to as an open or relaxed circle.

Fig. 28. Open circle.

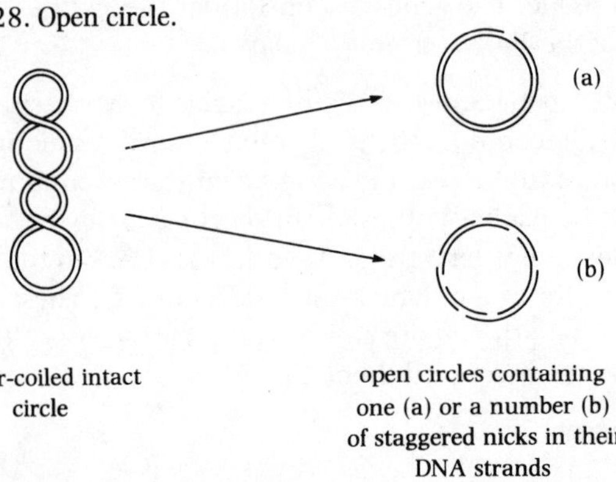

super-coiled intact circle

open circles containing one (a) or a number (b) of staggered nicks in their DNA strands

open reading-frame A reading-frame uninterrupted by stop codons. The existence of open reading-frames is usually inferred from the DNA (rather than the RNA) sequence.

operator The region of DNA preceding the coding sequence of a gene to which a repressor or activator will bind. The operator comes just after, or overlaps with, the promoter of a gene.

operon When the coding sequences of two or more genes are transcribed into a single mRNA molecule starting from a single promoter, the genes are said to form an operon. Many bacterial genes are organised into operons while this form of gene organisation is rare or absent in eukaryotes.

opine The general name given to rare amino acid and sugar derivatives found in crown-gall tumours.

ORF *See* open reading-frame.

ori The three-letter symbol for the origin of replication of a replicon.

origin The site on a DNA molecule at which DNA replication is initiated.

overlapping genes A situation discovered in certain small DNA viruses, e.g. the animal virus SV40, bacteriophages φX174 and G4 where a single nucleotide sequence encodes, from one strand, two functional proteins. This is possible if the sequence contains two open reading-frames in different phases, both of which can be recognised by the host cell's translation machinery.

The sequence below contains two AUG initiation codons in different phases, each of which defines a translatable reading-frame:

met   ala   asn   gly   pro   phe   thr   gln

A U G G C A A A U G G G C C U U U U A C A C A A

met   gly   leu   leu   his

P

packaging The act of forming a viral head particle around a nucleic acid molecule or of filling a preformed virus with nucleic acid. (*See in vitro packaging*)

PAGE *See* polyacrylamide gel electrophoresis.

palindromic sequence A bilaterally symmetrical DNA sequence which, therefore, reads the same in both directions. The target sites of most restriction enzymes are palindromic sequences. E.g.

```
            Point of symmetry
                    |
     A T T G C C G T T A
     T A A C G G C A A T
                    |
```

papovavirus A group of double-stranded DNA animal tumour viruses which includes SV40 and bovine papilloma virus. These two viruses form the basis of vector systems for mammalian gene cloning experiments.

par The genetic notation for the partition function. The locus *par* is found in some plasmids and is thought to enable plasmid molecules to be evenly distributed or partitioned between the two daughter cells when a bacterium divides. *par*$^+$ plasmids have a high stability.

Parafilm™ A stretchable film based on paraffin wax which is used to seal tubes and Petri dishes. Parafilm is a proprietary name which has come to stand for all products of that type.

partial digest, incomplete digest Treatment of a DNA sample with a restriction enzyme for a limited period so that only a proportion of the target sites in any individual molecule are cleaved. Partial digests are often performed with four base pair cutters to give an overlapping collection of DNA fragments for use in the construction of a gene bank.

pBR322 A widely used *E. coli* plasmid cloning vector of 4362 base pairs the complete sequence of which has been determined. pBR322 carries both tetracycline and ampicillin resistance genes. There are several restriction enzymes which have a single target site within one or the other resistance gene such that insertion of DNA at one site can be detected by insertional inactivation of that antibiotic resistance. pBR322 has a Col E1-type replication region and is present as a multicopy plasmid in exponentially growing cells at about 50 copies per chromosome. The copy number can be radically increased by chloramphenicol amplification. Because of its early appearance on the molecular biology scene and its versatility it has become established as the most widely used plasmid cloning vector.

Fig. 29. pBR322 (map).

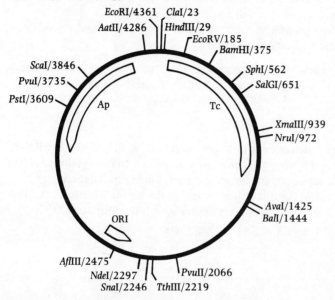

PEG, polyethylene glycol, carbowax This polymer is available in a range of molecular weights from *ca.* 1000 to *ca.* 6000. PEG 4000 and PEG 6000 are commonly used to promote cell or protoplast fusion, and to facilitate DNA uptake in the transformation of organisms such as yeast. PEG is also used to concentrate solutions by withdrawing water from them. The polymer has the general formula:

$$HOCH_2(CH_2OCH_2)_x CH_2OH$$

pEMBL, pEMBL8 (pronounced pembul) A class of cloning vectors constructed in the European Molecular Biology Laboratory in Heidelberg. pEMBL8 is designed to combine the advantages of plasmid cloning vectors with those of single-stranded DNA phages. This was achieved by inserting part of the phage f1 genome into a plasmid vector. The resulting construct behaves like a plasmid but is encapsidated into phage particles as single-stranded DNA when the host is superinfected with phage f1. This permits the easy preparation of single-stranded templates for the Sanger DNA sequencing procedure while retaining the simpler handling and high recombinant stability of a small plasmid.

penicillinase A type of β-lactamase.

periplasmic space, periplasm The space between the cytoplasmic membrane of a bacterium or fungus and its outer wall or membrane. The periplasm contains a number of secreted enzymes and other proteins; these are concerned with the binding and uptake of molecules from the growth medium, the degradation of high-molecular-weight substrates and the synthesis and organisation of the outer wall or membrane.

phage An abbreviation for bacteriophage, a virus which infects bacteria. The full word is, in fact, rarely used. (For specific phages *see* λ, μ, ϕX174)

phage λ *See* λ

phase, frame Nucleotide triplets are 'in phase' with one another if they can be read in groups of three from an AUG initiation codon and can, therefore, act as codons for amino acids during translation.

In the following nucleotide sequence the first UAG (stop codon) is out of phase but the second is in phase with the initiating AUG:

A U G U U U A A A U U A G C C C C C C G A U A G

| | out of phase stop codon | in phase stop codon |

Also used to describe triplets in the coding strand of DNA.

phasing *See* nucleosome phasing.

phasmid A hybrid molecule formed *in vivo* between a plasmid containing multiple *att* sites of λ and a bacteriophage λ derivative. The formation of a phasmid or its breakdown to release the plasmid and phage is controlled by the site-specific recombination system of λ. Phasmids can replicate as a plasmid (non-lytically) or as a phage (lytically). A number of sophisticated *in vivo* genetic manipulations can be accomplished using phasmids.

phenotype The observable or outward characteristics of an organism determined by its genotype and modulated by the environment.

ϕX174 (pronounced phi X one seven four) A bacteriophage which attacks *E. coli*. The nucleic acid within the viral particle is a single-stranded DNA circle of 5375 bases. The replicative form is a double-stranded circle and was the first complete DNA molecule to be sequenced. This sequence revealed the presence of overlapping genes.

physical containment The use of physical barriers to prevent the escape of genetically engineered organisms from the laboratory into the environment. Precautions employed include the maintenance of negative air pressure within the laboratory and the use of fume hoods with HEPA filters on their exhausts.

P_L, P_R Promoter leftward and promoter rightward, two strong promoters from bacteriophage λ. Transcription from both these promoters is under the control of the *cI* gene product.

plaque The 'hole' formed in a lawn of bacteria due to a lytic infection with a bacteriophage. A plaque may be formed from an original infection of a single cell with one phage particle. The release of progeny phage on cell lysis sets up a chain reaction of infection which permits the radial spread of the plaque. Virulent phages form clear plaques but temperate phages form turbid plaques due to the survival and division of those bacteria in which the phage has entered the lysogenic or prophage state.

plaque hybridisation, Benton–Davis technique, phage lift, plaque lift An adaptation of the technique of Southern hybridisation. Bacteriophage plaques on an agar plate are transferred to a nitrocellulose filter, or other suitable DNA-binding matrix, by pressing the filter gently onto the surface of the agar plate. A small amount of the phage and unpackaged phage DNA binds to the filter leaving behind the original plaques. The phage DNA is denatured on the filter which is then baked to attach the DNA firmly. The bound DNA can now be hybridised with a probe, usually ^{32}P-labelled DNA or RNA. Any homology between the probe and the DNA on the filter will allow the two species to anneal and the hybrids formed can be detected by autoradiography.

plasmid An extrachromosomal element capable of independent replication. A plasmid is not physically linked to the chromosome and thus can be lost from its host cell. Plasmids often carry genes which are not part of the central metabolism of the cell, e.g. antibiotic resistance, bacteriocin production, catabolism of hydrocarbons or plant tumour induction. Plasmids vary in size from below 1 kb to over 300 kb and vary in copy number from 1 to over 100 per chromosome. Many plasmids carry genes which mediate their own transfer (*tra* genes). Some also carry genes which enable them to mobilise segments of the host's chromosome from one cell to another (*cma* genes). Many cloning vectors are based on plasmids which have been manipulated to have desirable features such as antibiotic resistance genes, multiple copies, single restriction enzyme sites and strong promoters. The majority of plasmids are circular, double-stranded DNA molecules although recently some examples of linear plasmids have been discovered. (*See* episome)

plasmid rescue *See* homologous helper plasmid.

pMB9 A small multicopy plasmid which carries a tetracycline-resistance gene. It was used in the construction of the cloning vector pBR322. It has a Col E1 type replication region and can undergo chloramphenicol amplification. At one time pMB9 was used as a cloning vector but it has now been superseded by pBR322.

pock formation This phenomenon is seen in *Streptomyces*. A zone of inhibition of sporulation and retardation of growth of a plasmid-free strain is formed around a colony of a plasmid-containing strain. Some Streptomycete cloning vectors have pock formation as a phenotypic marker for plasmid containing strains. Pock formation is also known as lethal zygosis.

pol The gene which encodes reverse transcriptase in certain eukaryotic RNA viruses.

polish To convert single-stranded tails on the ends of linear DNA molecules into complete duplexes using either Klenow polymerase or T4 DNA polymerase. Such single-strand extensions may be generated either by the action of restriction enzymes or by shearing or sonicating DNA into random fragments.

pol **I, polymerase I** *See* DNA polymerase I.

poly(A) Polyadenylic acid. (*See* poly(A) tail)

polyacrylamide gels Often referred to, incorrectly, as acrylamide gels. These gels are made by cross-linking acrylamide with N,N′-methylene-*bis*-acrylamide. Polyacrylamide gels are used for the electrophoretic separation of proteins and also RNA molecules. DNA molecules usually have too high a molecular weight to migrate far in polyacrylamide.
   Polyacrylamide beads are also used as molecular sieves in gel chromatography and are marketed under the brand name 'Bio-gel'.

polyacrylamide gel electrophoresis, PAGE A method for separating nucleic acid or protein molecules according to their molecular size. The molecules migrate through the inert gel matrix under the influence of an electric field. In the case of protein PAGE, detergents such as sodium dodecyl sulphate are often added to ensure that all molecules have a uniform charge.
   Secondary structure can often lead to the anomalous migration of

molecules. It is common, therefore, to denature protein samples by boiling them prior to PAGE. In the case of nucleic acids, denaturing agents such as formamide, urea or methyl mercuric hydroxide are often incorporated into the gel itself which may also be run at high temperature. PAGE is used to separate the products of DNA-sequencing reactions and the gels employed are highly denaturing since molecules differing in size by a single nucleotide must be resolved.

poly(A) polymerase The enzyme which adds the poly(A) tail to nascent mRNA molecules. Most eukaryotes have two poly(A) polymerases; one in the nucleus which adds an oligo(A) stretch to the transcript, and another in the cytoplasm which extends the tail to its full length.

poly(A) tail The run of adenylic acid residues which is added to the 3′ end of many eukaryotic mRNA molecules following transcription. The length of the poly(A) tail varies from *ca.* 50 residues in *S. cerevisiae* to *ca.* 250 residues in mammalian cells. The function of the tail is unknown but it is thought to be involved in mRNA stability. Some bacterial mRNAs also have a poly(A) tail but it appears to be short-lived.

poly(dA) (polydeoxyadenylic acid) tails are added to the 3′ end of restriction fragments in the A's and T's method of gene cloning.

polymerase An enzyme which will form oligomeric molecules from monomers. A DNA polymerase will synthesise DNA from deoxynucleoside triphosphates using a complementary DNA strand and a primer. An RNA polymerase will synthesise RNA from monoribonucleoside triphosphates and a complementary DNA strand. (*See* reverse transcriptase)

polymerase I *See* DNA polymerase I.

polynucleotide A long chain of nucleotides in which each nucleotide is linked by a single phosphodiester bond to the next nucleotide in the chain. They can be double- or single-stranded and used to describe DNA or RNA.

polynucleotide kinase An enzyme, encoded by bacteriophage T4, which adds a phosphate to a 5′-hydroxyl group on either single- or double-stranded DNA and RNA molecules. One use of the enzyme is to end-label DNA restriction fragments, e.g. for sequencing by the Maxam–Gilbert method. The 5′ phosphate groups on the fragment are first removed using either bacterial alkaline phosphatase or calf intestinal alkaline phosphatase and replaced with radioactive phosphate using polynucleotide kinase and γ-^{32}P-ATP.

poly(U) glass fibre This is polyuridylic acid cross-linked to glass fibre and is used in the isolation of poly(A) mRNA from eukaryotic cells by affinity chromatography. The poly(A) tail on the mRNA molecule is hybridised to the poly(U) bound to the filter; the mRNA so isolated is later released under denaturing conditions. (*See* oligo-dT cellulose)

poly(U) sepharose This is polyuridylic acid covalently bound to agarose beads. It is used in the preparation of poly(A) mRNA from eukaryotic cells by affinity chromatography. The poly(A)-tail in the mRNA molecules is hybridised to the complementary poly(U) sequence bound to the beads and is thus purified away from other RNA molecules. The poly(A) mRNA may later be released from the poly(U)-sepharose column by washing with a denaturing buffer solution. (*See* oligo-dT cellulose, poly(U)-glass fibre)

positive control In genetics this refers to a method of control on the transcription of a gene or operon. In positive control a protein is required for transcription to take place. This protein, or positive regulator, binds to a nucleotide sequence upstream of the -35 sequence to which RNA polymerase binds and somehow interacts with the DNA and/or the RNA polymerase to enable transcription to start. The positive control protein often has to bind some small molecule before it can bind to DNA and activate transcription. In enzymology, positive control means that an enzyme needs to bind an effector molecule to be activated.

positive selection The situation where cells carrying a certain gene can be detected because the activity of that gene is essential for cell growth under certain conditions; e.g. antibiotic resistance genes are positively selectable.

Positive selection vectors have been designed such that only those which contain an inserted foreign DNA fragment will allow the cell to survive. Some examples are insertion into genes which are lethal to the host, i.e. restriction enzyme genes or bacteriophage lysis genes. In a cloning experiment using a positive selection vector all colonies surviving at the end of the experiment must contain a DNA insert.

power pack A transformer which regulates the supply of direct current electricity to an electrophoresis or isoelectric-focussing apparatus. Power packs can be regulated such that either the current, voltage or power supplied to the system is kept constant.

Pribnow box A nucleotide sequence found in prokaryotic promoters *ca.* 10 bp upstream from the transcriptional start site. It has the consensus sequence:

5′	T	A	T	A	A	T	G	3′
3′	A	T	A	T	T	A	C	5′
	-12	-11	-10	-9	-8	-7	-6	

(For eukaryotes, *see* TATA box)

primer DNA polymerase, unlike RNA polymerase, is unable to initiate the *de novo* synthesis of a polynucleotide chain. It can only add nucleotides to a free 3′ hydroxyl group at the end of a pre-existing chain. A short oligonucleotide, known as a primer, is therefore needed to supply such a hydroxyl group for the initiation of DNA synthesis. An RNA primer is used for the initiation of DNA replication *in vivo*. In the dideoxy (Sanger) DNA-sequencing procedure a synthetic DNA primer, complementary to some part of the single-stranded sequence of phage M13, is used to initiate DNA synthesis *in vitro*.

probe As a noun, a probe is a specific DNA or RNA sequence which has been radioactively labelled to a high specific activity. Probes are used to detect complementary sequences by hybridisation techniques such as Southern or Northern blotting or colony hybridisation. As a verb, 'to probe' is the act of hybridisation to detect a specific gene or transcript. E.g. 'We probed our bank with labelled rRNA to detect clones containing rDNA sequences'.

prokaryote, prokaryotic Organisms in which the genome is a single, circular DNA molecule which is free in the cytoplasm and not separated from the rest of the cell inside a membrane-bound nucleus. Within the genome, genes encoding proteins of related function are often clustered and several genes may be arranged in a coordinately regulated block called an operon. The prokaryotes are divided into the true bacteria (eubacteria), the archaebacteria (an ancient group which includes many salt-tolerant organisms) and the blue-green bacteria (sometimes called the blue-green algae, this is a group of photosynthetic bacteria which use the pigment phycocyanin in converting light energy to chemical energy).

promiscuous A term used to describe a broad host-range transmissible plasmid.

promoter The DNA region, usually upstream to the coding sequence of a gene or operon, which binds RNA polymerase and directs the enzyme to the correct transcriptional start site.

prophage The bacteriophage genome in its repressed state during lysogeny. Prophages may have either a chromosomal or an episomal location.

protease An enzyme which hydrolyses protein molecules.

protein A A protein found on the surface of certain strains of *Staphylococcus aureus* cells. It binds very tightly to the Fc portion of an antibody molecule only when the antibody is bound to an antigen. Protein A derivatives labelled with a fluorescent molecule or radioactive atom can be used in a variety of immunological screening techniques.

protein blot *See* Western blot.

protoplast A microbial or plant cell which has had its wall removed, usually by the action of lytic enzymes. Protoplasts are used to create hybrid cells via fusion and in some organisms, e.g. yeast and *Streptomyces*, protoplasts are used to facilitate transformation.

protoplast fusion A technique for producing hybrids between two cells which would not normally mate. The two cells may belong to the same or different species. The cell walls are removed from the two parent cells to create protoplasts and then fusion of the two cell membranes is promoted, usually by the addition of PEG (polyethylene glycol) and Ca^{2+} ions. These fusogenic agents cause proteins to migrate from certain regions of the two cell membranes and allow areas of naked phospholipid to fuse. Subsequent regeneration of the cell wall allows the propagation of the hybrid organism. If nuclear fusion does not follow cell fusion then heterokaryons, rather than diploid organisms, are produced. In some crosses the genetic contribution of the two parents to the stable hybrid can be markedly unequal.

pSC101 A small, non-conjugative plasmid which encodes tetracycline resistance. Its origin is not clear as it is now thought to have come from a contaminant in a transformation experiment. It was one of the first vectors used in genetic engineering. The pSC101 tetracycline-resistance gene was used in the construction of pBR322.

pseudogene A copy of a functional gene which is not itself transcribed. Pseudogenes are thought to originate from the integration into the genome of cDNA copies synthesised from mRNA molecules by reverse transcriptase. Pseudogenes therefore have no promoter and have a poly(dA) sequence at their 5′ ends. Because they are not subject to any evolutionary pressure to maintain their coding potential pseudogenes accumulate mutations and often have stop codons in all three reading frames.

Pseudomonas A genus of Gram-negative, rod-shaped bacteria which are widespread in nature and have the ability to degrade a large number of recalcitrant organic compounds. Genes encoding these degradative enzymes may be carried on plasmids. *Pseudomonas aeruginosa* is an opportunist pathogen of Man.

Pst I (pronounced pisst one) A type II restriction enzyme from the bacterium *Providencia stuartii* 164 which recognises the DNA sequence shown below and cuts it at the sites indicated by the arrows:

$$\downarrow$$
5′ C T G C A G 3′
3′ G A C G T C 5′
$$\uparrow$$

The cloning vector pBR322 has a single *Pst* I site within the ampicillin-resistance gene. The 3′ single-stranded extensions or sticky ends produced by *Pst* I are an ideal substrate for homopolymer tailing using terminal transferase. (A complete list of restriction enzymes can be found in Appendix 1)

Pu Purine. (*See* base)

Py Pyrimidine. (*See* base)

Q

Qβ A bacteriophage which infects *E. coli*. It contains a single-stranded circular RNA genome. The viral RNA is termed the + strand because it can function as an mRNA molecule and direct synthesis of viral proteins. The + strand also acts as a template for the synthesis of a complementary (−) strand. The resulting double-stranded molecule together with the Qβ replicase directs the synthesis of a large number of + strands. This autocatalytic replication can be harnessed to produce large quantities of a desired RNA molecule using recombinant RNA technology.

R

rabbit reticulocyte system A cell-free system prepared from lysed rabbit reticulocytes which is able faithfully to translate eukaryotic mRNAs from a wide variety of heterologous sources.

RAC (pronounced rack) Recombinant DNA Advisory Committee. A committee of the United States' NIH which formulates guidelines governing the performance of genetic engineering experiments.

radioimmunoassay An assay system which uses the specificity of the antibody–antigen reaction of an immunoassay method with the sensitivity of detection given by radioactive labelling. The radioactive atom employed is often ^{125}I and it is attached to the antibody by the enzyme horseradish peroxidase. Any antibody–antigen complexes are detected by autoradiography.

rDNA, ribosomal DNA The genes encoding ribosomal RNA. There are usually many copies of these genes and in eukaryotes they may be extrachromosomally located.

In some popular texts and journals rDNA is used as an abbreviation for recombinant DNA.

reading-frame mRNA molecules have their nucleotide sequence translated in groups of three nucleotides (called codons) by the ribosome. Each codon is represented by a single amino acid in the protein synthesised. The reading-frame defines which sets of three nucleotides are read as codons; this is determined by the initiation codon AUG, e.g. the sequence A U G G C A A A A U U U C C C would read as

AUG/GCA/AAA/UUU/CCC/ and not as
A/UGC/CAA/AAU/UUC/CC

(*See* open reading-frame)

readthrough In the presence of a suppressor-tRNA, an amino acid can be inserted into a growing polypeptide chain in response to a stop codon. Termination is thus prevented and a longer than usual polypeptide synthesised. This process is known as readthrough and the protein product is a readthrough protein.

recA The genetic notation for a gene which encodes a protein involved in homologous recombination in *E. coli*. The *recA* gene product also acts as a protease which specifically degrades certain proteins such as that encoded by the *cI* gene of phage λ. Cells containing a mutant *recA* gene are often used when fragments of the *E. coli.* chromosome are cloned in *E. coli*. This prevents homologous recombination between the chromosomal gene and its cloned copy.

recBC The genetic notation for an *E. coli* gene whose product mediates recombination. The enzyme encoded by the *recBC* gene is an ATP-dependent nuclease specific for double-stranded DNA. This enzyme will degrade linear DNA that has been taken up by transformation.

recircularization The ligation of the two ends of a linear DNA molecule to reform a circular plasmid. Recircularization is favoured by low DNA concentrations since, under such conditions, the two ends of the same DNA molecule are likely to be in closer proximity to each other than to the ends of other DNA molecules. In cloning experiments it is often necessary to take steps to prevent the recircularization of the vector, for instance by removing 5'-phosphate groups with alkaline phosphatase.

recognition site The nucleotide sequence to which a restriction enzyme binds initially. For type II restriction enzymes (those used in gene-cloning experiments) it is also the sequence within which the enzyme specifically cuts the DNA i.e. for type II enzymes, the recognition site and the target site are the same sequence. Type I enzymes, however, bind to their recognition site and then cleave the DNA at some more or less random position outside that recognition site. It is this random property which disqualifies type I enzymes from use as tools in recombinant DNA technology.

recombinant A term used in both (i) classical genetics and (ii) *in vitro* genetics.
(i) An organism containing a different combination of alleles from either of its parents, e.g. Parents: AABB and aabb; recombinant: AaBb. Such an individual may result from crossing-over events or from the independent assortment of different chromosomes at meiosis.
(ii) A molecule containing a new combination of DNA sequences. The word is also used as an adjective, e.g. recombinant DNA.

recombinant DNA DNA molecules in which sequences which are not naturally contiguous have been placed next to each other by *in vitro* manipulations. The different sequences within a recombinant DNA molecule will frequently have come from entirely different organisms.

recombinant RNA A term used to describe RNA molecules joined *in vitro* by T4 RNA ligase. This technique may be used to join an RNA sequence to a phage Qβ replicase template. The recombinant RNA molecule can be autocatalytically replicated by Qβ replicase to produce large amounts of an RNA sequence of interest.

reiterated Nucleotide sequences which occur many times within a genome are said to be reiterated. In lower organisms the only highly reiterated sequences are usually the rRNA genes, but in higher organisms there is a lot of highly reiterated DNA which can be recognised by its rapid renaturation kinetics.

relaxed circle *See* open circle.

relaxed plasmid A high copy number plasmid, usually represented by more than 20 copies per chromosome. Although the term 'relaxed' refers to the relaxed control of plasmid replication this is somewhat misleading. In fact, these multicopy plasmids do control their copy number but set it at a higher value than do low copy number plasmids.

renaturation The process in which separated complementary strands of nucleic acid reform base pairs to form a double-stranded structure after denaturation. Some simple proteins can also be renatured and regain their function.

repeat A nucleotide sequence which occurs more than once in a DNA molecule. Repeat sequences may be in the same (direct repeats) or in the opposite (inverted repeats) orientation to one another.

replacement, gene replacement A method of substituting a cloned gene, or part of a gene, which may have been mutated *in vitro*, for the wild-type copy of the gene within the host's chromosome. (For details, *see* transplacement, homogenotization)

replacement vector A phage cloning vector in which a portion of the phage genome is replaced by the foreign DNA fragment to be cloned.
  Replacement vectors are usually based on phage λ. The portion of the λ genome which is replaced (known as the stuffer fragment) lies within the central region of the λ DNA which is not required for lytic growth. Cloning protocols which use replacement vectors frequently involve some means of preventing the religation of the stuffer fragment and/or identifying the clones in which it is replaced by foreign DNA. In some vectors, the stuffer fragment contains the gene for β-galactosidase. In this case non-recombinant phage will give blue plaques on X-gal plates whereas recombinant phage (lacking the β-galactosidase gene) will give white plaques. In other vectors the presence of the stuffer fragment prevents a lytic infection of certain strains of host bacteria.

replication The duplication of genomic DNA or RNA as part of the reproductive cycle of a cell or virus.

replicative form *See* RF.

replicator A DNA segment which contains an origin of replication and is able to promote the replication of a plasmid DNA molecule in a host cell. (*See ARS*)

replicon That portion of a DNA molecule which is replicable from a single origin. Plasmids and the chromosomes of bacteria, phages and other viruses usually have a single origin of replication and, in these cases, the entire DNA molecule constitutes a single replicon. Eukaryotic chromosomes have multiple internal origins and thus contain several replicons.

    The word is often used in the sense of a DNA molecule capable of independent replication, e.g. 'The shuttle vector pJDB219 is a replicon in both yeast and *E. coli*'.

repressor A protein which binds to a specific DNA sequence (the operator) upstream from the transcription initiation site of a gene or operon and prevents RNA polymerase from commencing mRNA synthesis. Examples of repressors are the *cI* protein of bacteriophage λ and the *lacI* protein of the *lac* operon.

resistance gene A gene which determines resistance to some inhibitor or toxin, e.g. an antibiotic or heavy metal. Resistance genes are frequently included in vector molecules since they enable the selection of organisms containing the vector.

res⁻ mod⁻ See *r⁻ m⁻*.

res⁻ mod⁺ See *r⁻ m⁺*.

restriction The process by which certain bacteria destroy the infecting DNA of bacteriophages using site-specific endonucleases. The host range of the phage is thereby restricted while host DNA is protected from degradation by post-synthetic modification.

restriction enzyme An endonuclease which recognises a specific sequence of bases within double-stranded DNA. Type I restriction enzymes bind to this recognition site but subsequently cut the DNA at approximately random sites. Type II restriction enzymes both bind and cut within their recognition or target site. They may make the cuts in the two DNA strands exactly opposite one another and generate blunt ends or they may make staggered cuts to generate sticky ends. It is the type II restriction enzymes which have been exploited in recombinant DNA technology. (*See* HCRM) (Specific restriction enzymes are listed by name and also in Appendix 1)

restriction fragment *See* fragment.

restriction site The specific nucleotide sequence in DNA recognised by a type II restriction endonuclease and within which it makes a double-stranded cut. Restriction sites usually comprise four or six base pairs and have bilateral symmetry, e.g.

G A G¦C T C
C T C¦G A G

The two strands may be cut either opposite to one another, to create blunt ends, or in a staggered manner giving sticky ends, depending on the enzyme involved. (*See* restriction enzyme, HCRM.) (Individual enzymes are listed by name and in Appendix 1)

retrovirus A class of eukaryotic RNA viruses which can integrate a DNA copy of their genome into the chromosome of their host. The DNA copy is synthesised by a viral enzyme called RNA-dependent DNA polymerase or reverse transcriptase. Retroviruses, in the integrated state, have long, terminal direct repeat sequences and appear to integrate into the host's DNA in a transposon-like manner. These repeats contain enhancer and promoter elements. Some retroviruses have been adapted as vectors for mammalian gene cloning.

reverse genetics The use of recombinant DNA techniques to investigate gene function, usually by introducing specific mutations *in vitro*, e.g. by site-directed mutagenesis, and then returning the mutated gene to its original host, and often to its original site in that host, by homogenotization or transplacement. It is the opposite of conventional genetics where mutations are introduced at random into the whole genome followed by a lengthy procedure to select for the desired mutant.

reverse transcriptase, RNA-dependent DNA polymerase An enzyme found in retroviruses (e.g. avian myeloblastosis virus, AMV) which makes a DNA copy of an RNA molecule when primed with a suitable oligodeoxyribonucleotide. The enzyme is commonly used to make DNA copies of mRNA molecules in cDNA cloning.

RF, replicative form The intracellular form of viral nucleic acid which is active in replication, e.g. M13 phage particles contain a single-stranded DNA circle while the RF of the same molecule is double-stranded.

Rhizobium A genus of soil bacteria which can set up a symbiotic association with the roots of leguminous plants, e.g. pea, clover. *Rhizobium* bacteria are found in special nodules on the roots of the plants and within the bacteria nitrogen fixation takes place. The plants are provided with NH_3 and the bacteria obtain carbon compounds for growth from the plant. Many, if not all, Rhizobia contain one or more large plasmids some of which give the bacterium the ability to colonise a particular species of plant.

ribonucleoside *See* nucleoside.

ribonucleotide *See* nucleotide.

ribosomal RNA *See* rRNA.

ribosome The subcellular complex responsible for protein synthesis. It consists of two subunits which both contain one or more RNA molecules (rRNAs, ribosomal RNAs) and a large number of proteins. The small subunit is responsible for binding mRNA and is subsequently joined by the large subunit which accepts the aminoacyl tRNA molecule and carries out the process of peptide bond formation.

ribosome-binding site, Shine–Dalgarno sequence, S–D sequence A particular sequence of nucleotides in an mRNA molecule to which a ribosome will bind. Such a sequence is complementary to the 3′ end of the 16s rRNA. Ribosome-binding sites are 3–9 bases long and precede the translational start codon by 3–12 bases. The relationship between the S–D sequence for *E. coli* and the 3′ end of the 16s rRNA is shown below:

consensus S–D sequence

5′   A A G G A G G U   3′ mRNA

3′   A U U C C U C C A C U A G   5′ 3′ end of *E. coli*
16s ribosomal RNA

Named after the Australian researchers who first identified it.

ribozyme An RNA sequence with enzyme-like activity. It is an internal guide sequence (IGS) contained within an intron of a eukaryotic transcript which is able to catalyse the splicing of that intron *in vitro* in the absence of any enzyme or other protein molecule. Ribozyme activity was first demonstrated with the intron of the pre-rRNA of the protozoan *Tetrahymena*.

Ri plasmid A class of large conjugative plasmids found in the soil bacterium *Agrobacterium rhizogenes*. Ri plasmids are responsible for hairy root disease of certain plants. A segment of the Ri plasmid is found in the genome of tumour tissue from plants with hairy root disease.

R-looping The technique in which an RNA molecule is annealed to the complementary strand of a partially denatured DNA molecule. The formation of the RNA·DNA hybrid displaces the opposite DNA strand as a single-stranded bubble. These R-loops can be visualised under the electron microscope using the Kleinschmidt technique.

It was the R-looping technique which first revealed the presence of introns in eukaryotic genes. (*See* D-looping)

r^-m^- Shorthand genetic nomenclature for a bacterial strain deficient in both the restriction and modification of DNA.

r^-m^+ Shorthand genetic nomenclature for a bacterial strain deficient in the restriction but not the modification of DNA. Such strains are useful hosts for the cloning of heterologous genes.
(r^+m^- strains would commit suicide by restricting their own, unmodified, DNA)

RNA, ribonucleic acid The alternate reservoir of genetic information to DNA. A number of viruses have single-stranded or double-stranded RNA genomes. In organisms, RNA is used as a primer in DNA replication and is essential to the expression of the genetic information contained within the DNA. RNA differs from DNA in having ribose instead of deoxyribose as the sugar moiety in its nucleotides and in having uracil instead of thymine as one of its two pyrimidine bases. RNA, but not DNA, may be degraded by alkaline hydrolysis. (*See* alkaline hydrolysis, base, exon, guide sequence, HnRNA, intron, mRNA, Northern blot, poly(A) tail, recombinant RNA, rRNA, tRNA)

RNA ligase An enzyme which can join RNA molecules together. (*See* T4 RNA ligase)

RNA polymerase The enzyme which transcribes DNA into RNA. It is able to initiate RNA synthesis on a DNA template in the absence of any primer molecule. Prokaryotes have a single RNA polymerase which synthesises all classes of RNA molecules. Eukaryotes have three RNA polymerases with different transcriptional specificities: RNA polymerase A (or I) synthesises the large rRNA precursor; B (or II) synthesises mRNA; C (or III) synthesises tRNA and 5s rRNA species.

RNase, ribonuclease An enzyme which hydrolyses RNA. A wide range of ribonucleases are used in the characterisation and sequencing of RNA molecules. Single-strand specific, e.g. RNase T1, and double -strand specific, e.g. RNase III, enzymes exist and many are involved in the *in vivo* processing of RNA precursor molecules.

rolling circle A DNA replication mechanism. A circular double-stranded DNA molecule, e.g. the genome of phage λ is nicked in one strand and one of the nicked ends peeled back. The exposed single-stranded regions are replicated and this drives the further displacement of the peeled-back strand. Eventually the replicating fork will come back to the origin. Replication may either stop there or continue to produce a long string of concatemeric DNA.

Fig. 30. Rolling circle.

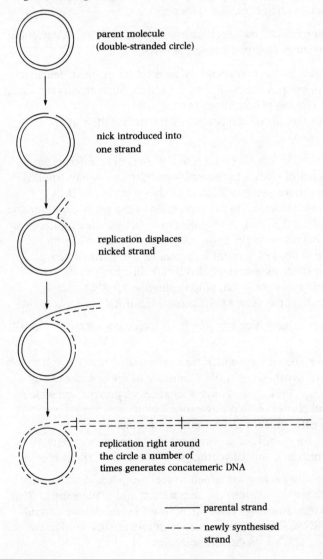

parent molecule
(double-stranded circle)

nick introduced into
one strand

replication displaces
nicked strand

replication right around
the circle a number of
times generates concatemeric DNA

———— parental strand

– – – – newly synthesised
strand

RP4 A promiscuous antibiotic-resistance plasmid first isolated from *Pseudomonas aeruginosa*. Several plasmids identical to RP4 were isolated around the world at approximately the same time, e.g. RP1, R68, R1822, RK2. RP4 is the prototype of the Inc-P1 group of plasmids, of great interest because of their broad host-range. Several derivatives of RK2 and RP4 have been made for use as broad host-range cloning vectors.

rRNA, ribosomal RNA The RNA molecules which are essential structural and functional components of ribosomes, the organelles responsible for protein synthesis. The different rRNA molecules are known by their sedimentation (Svedberg, S) values. *E. coli* ribosomes contain one 16S rRNA molecule (1541 nucleotides long) in the small subunit and a 23S rRNA (2904 nucleotides) and a 5S rRNA (120 nucleotides) in the large subunit. These three rRNA molecules are synthesised as part of a large precursor molecule which also contains the sequences of a number of tRNAs. Special processing enzymes cleave this large precursor to generate the functional moieties.

Eukaryotic cytoplasmic ribosomes contain four rRNA molecules. The small subunit contains a single 18S rRNA, the large subunit contains a 28S rRNA, a 5S rRNA and a 5.8S rRNA. The exact sizes of these molecules vary from species to species. In the large subunit, the 5.8S rRNA is found tightly hydrogen bonded to the 28S species. The 5.8S, 28S and 18S species are synthesised as parts of a single precursor molecule of about 42S by RNA polymerase I. As in *E. coli*, processing enzymes release the mature molecules from this transcript. The 5S rRNA molecule is synthesised independently by RNA polymerase III and the genes encoding this species are often not linked to those specifying the large precursor.

RSF1010 A small 8.5 kb non-conjugative, but highly mobilisable, broad host-range plasmid of the incompatibility (Inc) Q group. It encodes Su^r and Sm^r and possesses several single restriction enzyme sites for cloning. Several derivatives which have different antibiotic-resistance genes and different restriction enzyme sites have been constructed in order to improve its versatility as a cloning vector. The plasmids NTP2 and R300B are identical to RSF1010.

runaway replication Uncontrolled replication of plasmid molecules which multiply within the host building up to several thousand copies. This usually halts cell division but is a very powerful way of amplifying a gene product. Plasmid vectors which are temperature sensitive for a copy number control element have been constructed as runaway replication vectors.

95

running buffer The buffer in which the electrodes of an electrophoresis system are immersed.

run out To separate nucleic acid or protein molecules by gel electrophoresis.

S

Saccharomyces cerevisiae The yeast which is used commercially in bread-making and in the production of alcoholic beverages and industrial alcohol. It is probably the best-characterised eukaryote in molecular biological terms. *S. cerevisiae* is able to divide vegetatively in either the haploid or the diploid phase. This permits the isolation of recessive mutations in haploids and complementation testing in diploids. Diploids can be induced to go through meiosis to give four haploid spores. The *S. cerevisiae* life cycle is shown below.

Fig. 31. *Saccharomyces cerevisiae* (life cycle).

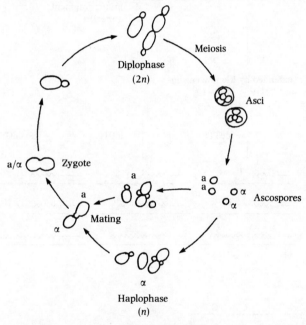

Transformation procedures which permit DNA uptake by either protoplasts or whole yeast cells have been developed and a wide range of cloning vectors constructed. (*See* YCp, YEp, YRp) (For the genetic map of *S. cerevisiae see* Appendix 4)

Sam An amber mutation in the S gene of phage λ. The S gene is involved with damaging the host cell membrane to initiate cell lysis. λ vectors often contain the *Sam* mutation as a means of biological containment. (*See* λ WES)

Sanger method, chain terminator technique, dideoxy technique A
method of sequencing DNA named after its main developer, F. Sanger.
The technique uses a single-stranded DNA template, a short DNA
primer and Klenow enzyme to synthesise a complementary DNA
strand. The primer is first annealed to the single-stranded template and
the reaction mixture is then split into four aliquots and
deoxynucleoside triphosphates (dNTPs) plus a dideoxynucleoside
triphosphate (ddNTP) are added such that each tube has a different
ddNTP. The Klenow enzyme will incorporate a ddNTP opposite its
complementary base on the template but no further dNTPs can be

Fig. 32. Sanger method.

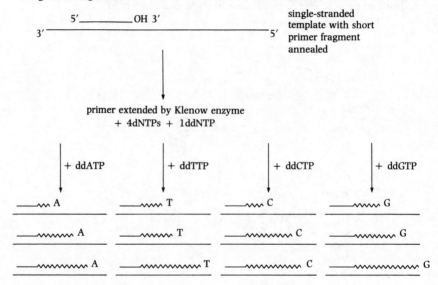

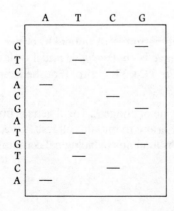

added as the ddNTP lacks a 3' hydroxyl group. The ratio of ddNTP to dNTP is such that the Klenow enzyme will terminate the growing DNA chain at all positions at which the ddNTP can be inserted and so a nested set of fragments is formed which all have one end, the primer, in common. One of the dNTPs in each reaction is labelled (usually it is ^{32}P-αdATP) so that when the four reaction mixtures are electrophoresed through a polyacrylamide gel and X-ray film is placed against the gel a band pattern or ladder is formed from which the DNA sequence can be read directly.

The single-stranded DNA template is often obtained by use of the M13 phage vectors.

Sau 3A A type II restriction enzyme from *Staphylococcus aureus* 3A which recognises the sequence

$$
\begin{array}{c}
\downarrow \\
5'\ \text{G A T C}\ 3' \\
3'\ \text{C T A G}\ 5' \\
\uparrow
\end{array}
$$

and cleaves at the sites denoted by the arrows. This four base pair target site will occur frequently in genomic DNA (on average, once every 4^4 or 256 base pairs). *Sau* 3A, therefore, is often used to partially digest the total genomic DNA of an organism to yield a semi-random, overlapping set of large DNA fragments in order to construct a gene bank. The GATC extension or sticky end generated by *Sau* 3A will hybridise to the sticky ends produced by cleavage of DNA by the enzymes *Bam* HI, *Bgl* II, *Bcl* I and *Xho* II. Thus any pairwise combination of ends generated by these enzymes can anneal and be joined by DNA ligase. *Sau* 3A partials are often inserted into a *Bam* HI site on a vector molecule. (A full list of restriction enzymes can be found in Appendix 1)

scaffolding A technique which enables any gene or DNA sequence to be integrated into the genome of a desired organism. It was originally described for *Bacillus subtilis* but is applicable to most organisms. First, a gene or small DNA sequence from the desired organism is cloned into a circular vector. This is then used to integrate the vector carrying that DNA sequence into the chromosome by homologous recombination. Subsequently any DNA sequence cloned using the same vector can be inserted into the chromosome by virtue of the homology between the transforming and the integrated vectors.

screen To survey a large number of individuals for a desired activity or phenotype, e.g. 'Amp-resistant transformants were selected and then screened for tetracycline sensitivity'.

A screen (noun) is such a survey.

screening *See* screen. Also a technique for mutant enrichment. E.g. in penicillin screening the mutagenised culture is placed under conditions in which wild-type, but not mutant, bacteria will grow. This culture is then treated with the antibiotic penicillin, which kills the growing (wild-type) cells but has no effect on the non-growing (mutant) cells. The population of viable bacteria is thus enriched for mutant cells.

S–D sequence *See* ribosome-binding site.

secrete To transfer a protein molecule through a membrane. Proteins may be secreted into (i) an intracellular compartment, e.g. a vacuole or mitochondrion, (ii) the periplasmic space, or (iii) the culture medium. Secretion out of the cell and into the culture medium is often described as excretion.

selection The establishment of a set of culture conditions under which organisms with a desired genotype will grow in preference to those with other genotypes. The geneticist must design appropriate culture conditions so that only organisms of the desired type can grow at all. (*See* positive selection)

self-ligation The joining of the two ends of a restriction fragment to each other using DNA ligase. The reaction is carried out at low DNA concentrations to reduce the likelihood of two molecules being ligated together. The procedure is often used in sub-cloning.

sense strand The strand of duplex DNA which is transcribed into a complementary mRNA (or other functional RNA) molecule.

```
                          sense strand
    DNA    5'  TACTTTCGCAAATCACCCGCGGGGATA   3'
           3'  ATGAAAGCGTTTAGTGGGCGCCCCTAT   5'
                        anti-sense strand

    RNA    5'  AUGAAAGCGUUUAGUGGGCGCCCCUAU   3'
```

sequencing The determination of the order of nucleotides in a DNA or RNA molecule or that of amino acids in a polypeptide chain. (For DNA sequencing, *see* Sanger method, Maxam–Gilbert method)

sequencing gel A long polyacrylamide slab gel which has sufficient resolving power to separate single-stranded fragments of DNA or RNA which differ in length by only a single nucleotide. Electrophoresis is carried out at high voltage and with the gel in a vertical position. Urea is usually included in the gel mixture as a denaturing agent. This prevents internal base pairing within the single-stranded molecules and ensures that their relative speed of migration is solely dependent on their length. Such gels are used to separate the radioactively labelled products of, for example, the Maxam–Gilbert or the Sanger sequencing reactions.

sex-factor A plasmid which promotes its own transfer via bacterial conjugation.

shear To fragment DNA molecules into smaller pieces. DNA, as a very long and fairly stiff molecule, is very susceptible to hydrodynamic shear forces. Forcing a DNA solution through a hypodermic needle will fragment it into smaller pieces. The size of the fragments obtained is inversely proportional to the diameter of the needle's bore. The actual sites at which the shear force breaks a DNA molecule are approximately random. Therefore DNA fragments may be generated by random shear and then cloned (by either tailing their ends or using linkers) so as to create a complete gene library of an organism. This method is little used now, having been replaced by the use of partial digests with four base pair cutters, e.g. *Sau* 3A as a means of generating random DNA fragments.

Shine–Dalgarno sequence, SD sequence *See* ribosome-binding site.

shot-gun A shot-gun experiment is one in which random fragments of an entire genome are cloned into a vector. A particular gene may then be selected for, or a gene bank established which is subsequently screened for sequences of interest. The word is also used as a verb, e.g. 'We shot-gunned *Xenopus* DNA into pBR322'.

shuttle vector, bifunctional vector A vector molecule which is able to replicate in two different host organisms and can therefore be used to 'shuttle' genes from one to the other. E.g. the YEp, pJDB219, is a shuttle vector able to replicate in *E. coli* from its pMB9 origin and in *S. cerevisiae* from its 2μ plasmid origin.

signal peptide *See* signal sequence.

signal sequence, signal peptide A short, 15–30 amino acid, segment at the N-terminus of a secreted or exported protein. This signal sequence is recognised by some part of the cell's protein-processing machinery and the protein is then secreted through the membrane of the cell or one of its organelles. The signal sequence is usually removed at some point in the secretion process by a specific protease and is therefore not present in the mature protein.

simian virus 40 A papovavirus which normally infects monkey (simian) cells. Like bacteriophage λ it is able either to establish a lytic infection or it may integrate into the host chromosome. In the latter case it transforms the cell into a tumourous state. SV40 has a small, circular double-stranded DNA genome with a molecular weight of 3×10^6. SV40 DNA and its derivatives have proved to be useful vector molecules in mammalian cloning systems.

single copy A gene or DNA sequence which occurs only once per (haploid) genome. Most structural genes, those encoding proteins, are single-copy genes.

single-stranded A term used to describe nucleic acid molecules consisting of only one polynucleotide chain. The genomes of certain male-specific phages, e.g. M13, are single-stranded DNA molecules. rRNA, mRNA and tRNA are all single-stranded nucleic acids, but they all contain double-stranded regions formed by the intra-strand base-pairing of self-complementary sequences.

site-specific A term used to describe any process or enzyme which acts at a defined sequence within a DNA or RNA molecule. Type II restriction enzymes are site-specific endonucleases and the recombination systems encoded by some transposons are site-specific, as is the integration of phage λ into the *E. coli* chromosome.

site-specific mutagenesis, site-directed mutagenesis, directed mutagenesis A technique in which a cloned gene is specifically mutated *in vitro* and then used to replace the wild-type copy of the gene in the donor organism. The technique has two advantages over conventional mutagenesis procedures: (i) The proportion of desired mutants obtained is high, up to 50% of the viable cells are mutant in some methods. (ii) No other genes in the organism are mutated and therefore all the desirable characteristics of the strain are retained.

In localised random mutagenesis, mutations are induced more or less at random within a limited portion of the gene. This involves introducing a small, single-strand gap at a specific position within the gene. Mutations are then introduced either by treating with bisulphite, which changes cytosine residues into uracils, or repairing the gap using some error-prone system, e.g. using reverse transcriptase instead of DNA polymerase or incubating Klenow polymerase with Mn^{2+} instead of Mg^{2+} ions in the buffer.

In oligonucleotide-directed mutagenesis, a synthetic oligonucleotide about 14 bases long is used to introduce pre-determined base changes into the gene. The oligonucleotide is annealed to a single-stranded copy of the recombinant molecule containing the gene. The fragment is then used to prime the synthesis of a second strand by Klenow polymerase. Following ligation the double-stranded molecule is transformed into the host where replication should result in 50% of cells containing the desired mutation.

102

six base-pair cutter A type II restriction enzyme which recognises a six base-pair DNA sequence and makes a double-stranded cut within it.

Sma I (pronounced smar one) A type II restriction enzyme from the bacterium *Serratia marcescens* which recognises the DNA sequence shown below and cuts it at the sites indicated by the arrows:

$$\downarrow$$
5′ C C C G G G 3′
3′ G G G C C C 5′
$$\uparrow$$

(A full list of restriction enzymes can be found in Appendix 1)

Sm^r *See* streptomycin resistance.

snap-back *See* fold-back.

snRNA, small nuclear RNA *See* guide sequence.

snurp, small nuclear ribonucleoprotein *See* guide sequence.

S1 nuclease A nuclease purified from the filamentous fungus *Aspergillus oryzae*. S1 degrades single-stranded RNA or DNA to 5′ mononucleotides. It is used in assessing the extent of a hybridisation reaction by removing unpaired regions. It is also used to remove the sticky ends of restriction fragments. In 'S1 mapping' the coding region of a gene is detected by performing mRNA–DNA hybridisation and removing unpaired DNA with S1.

Southern blot, Southern transfer A technique which combines the resolving power of agarose gel electrophoresis with the sensitivity of nucleic acid hybridisation. DNA fragments separated in an agarose gel are denatured *in situ* and then blotted or transferred, usually by capillary action, from the gel to a nitrocellulose sheet, or other binding matrix, placed directly on top of the gel. Single-stranded DNA binds to the nitrocellulose and is then available for hybridisation with labelled, ^{32}P or biotinylated, single-stranded DNA or RNA. The labelled nucleic acid is known as the probe and, in the case of DNA, is often prepared by nick translation. The hybrids are detected by autoradiography, in the case of ^{32}P, or a colour change, in the case of a biotinylated probe. A very sensitive and powerful technique, it is often described as 'blotting'. The technique is named after its inventor, E. M. Southern.

spheroplast, sphaeroplast A microbial or plant cell from which most of the cell wall has been removed, usually by enzymic treatment. Strictly, in a spheroplast, some of the wall remains, while in a protoplast the wall has been completely removed. In practice, the two words are often used interchangeably. (For applications, *see* protoplast)

splicing The mechanism by which intron sequences are removed from precursor RNA molecules and adjacent exon sequences are religated. Where enzymes are involved in this set of reactions they are often termed 'splicases'. 'Gene splicing' is often used as a general term for recombinant DNA technology. A foreign gene may be said to be spliced into a vector molecule.

split gene A eukaryotic gene in which the coding sequence is divided up by a number of non-coding regions called introns. (*See* exon, guide sequence, splicing)

Sst I A type II restriction enzyme produced by the bacterium *Streptomyces stanford* which recognises the DNA sequence shown below and cuts at the sites indicated by the arrows:

$$\downarrow$$
5′ G A G C T C 3′
3′ C T C G A G 5′
$$\uparrow$$

(A complete list of restriction enzymes can be found in Appendix 1)

stable This has its usual English meaning. Stable plasmids are ones which are lost only infrequently from their host cells even under non-selective conditions. In bacteria, plasmid stability appears to depend on two factors, copy number and the plasmid locus *par* which ensures the equitable partition of plasmid molecules into daughter cells at division.

Staphylococcus aureus A Gram-positive spherical bacterium which commonly inhabits the skin of humans. Several small antibiotic-resistant plasmids have been isolated from *S. aureus* and have formed the basis of many cloning vectors for *Bacillus* species particularly *Bacillus subtilis*.

start codon, initiator codon The trinucleotide in an mRNA molecule with which the ribosome starts the process of translation. The start codon sets the reading-frame for translation. The most commonly used start codon is AUG which is decoded as methionine in eukaryotes and as N-formylmethionine in prokaryotes. AUG appears to be the only start codon used by eukaryotes, while in bacteria GUG (valine) may sometimes be employed.

sticky ends, cohesive ends The single-stranded ends left on a restriction fragment by many type II restriction enzymes. These unpaired regions are available for hybridisation with complementary ends on other fragments during gene cloning experiments. (*See* 5′ extension, 3′ extension)

stop codon, termination codon A codon for which there is no corresponding tRNA molecule to insert an amino acid into the polypeptide chain. Protein synthesis is hence terminated and the completed polypeptide released from the ribosome. There are three stop codons: UAA (ochre), UAG (amber) and UGA (opal). Mutations which generate any of these three codons in a position which normally contained a codon specifying an amino acid are known as nonsense mutations. Stop codons can also be called nonsense codons. (*See* suppressor)

streptavidin A protein from *Streptomyces avidinii* which has a very high affinity for the vitamin biotin. It is used as part of a complex with biotin and horseradish peroxidase to detect biotinylated DNA hybrids as a non-radioactive alternative in a Southern transfer experiment.

Streptomycetes A group of Gram-positive sporulating bacteria which grow in a mycelial (filamentous) manner. They are very common soil organisms and are involved in the breakdown and recycling of many polymeric carbon compounds, e.g. cellulose, protein and starch. Streptomycetes produce over 60% of the known antibiotics and because of this much attention has been devoted to these organisms by the pharmaceutical industry.

streptomycin An antibiotic which inhibits the elongation step of protein synthesis. It is produced by several strains of *Streptomyces* and is widely used in laboratories since there are many streptomycin-resistant strains available.

streptomycin resistant, Sm^r; streptomycin sensitive, Sm^s Resistance or sensitivity to the inhibitory effects of the antibiotic streptomycin. The resistance can be of two major types: (i) alteration of the target site on one of the ribosomal proteins, and (ii) inactivation of the antibiotic by phosphorylation, adenylation or acetylation. Many plasmids carry a gene for Sm^r which encodes an inactivating enzyme.

stringent plasmid A plasmid whose copy number is under strict control and therefore has only one or two copies per chromosome, i.e. a low copy number plasmid, e.g. the F plasmid.

strong promoter A promoter for which RNA polymerase has a high affinity and which will direct the synthesis of large amounts of mRNA.

stuffer fragment A non-essential fragment of a λ phage cloning vector which is replaced by the DNA fragment to be cloned. (*See* replacement vector)

sub-clone A method in which smaller DNA fragments are cloned from a large insert which has already been cloned in a vector.

sucrose gradient A density gradient which is preformed by mixing sucrose solutions of different concentrations, e.g. 10 and 40%, so that there is a linear increase in concentration from the top to the bottom of the tube. Sucrose gradients are commonly used to separate molecules or particles on the basis of their sedimentation coefficients, which depend on their size and shape rather than their density. Sucrose gradient centrifugation therefore separates particles according to their rate of migration down the centrifuge tube. Unlike CsCl gradient centrifugation it is not an equilibrium technique; if centrifugation is continued for long enough all the particles will be pelleted to the bottom of the tube.

super-coiled One of the conformations a DNA molecule can adopt. A circular plasmid is super-coiled when one of its strands has been underwound or overwound in relation to the other. If the strands are sealed again the molecule will be under torsional strain and will coil onto itself in the characteristic shape of a super-coiled molecule which is similar to a wound-up elastic band. All plasmids and large domains of bacterial chromosomes are super-coiled, usually by the action of the enzyme DNA gyrase. (*See* intercalating agent, ethidium bromide)

superinfecting phage A bacteriophage which infects a cell which already harbours the same type of bacteriophage.

suppressor Mutations in suppressor genes are able to overcome (suppress) the effects of mutations in other, unlinked, genes. A common, and very useful, kind of suppressor mutation occurs within the gene encoding a tRNA molecule and results in a change in the tRNA's anticodon. Such a mutant tRNA can reverse the effects of chain-terminating mutations, such as amber or ochre, in protein-encoding genes. It does this by permitting the insertion of an amino acid in the growing polypeptide chain in response to the stop codon in the mRNA molecule. Thus, premature chain termination is prevented and the protein synthesised retains at least partial function.

surrogate genetics *See* reverse genetics.

Su^s, Su^r Sulphonamide sensitive, sulphonamide resistant.

SV40 *See* simian virus 40.

SVGT5 A mammalian cell cloning vector derived from SV40 by the removal of a *Hind* III – *Bam* HI fragment. This fragment contains the coding sequence for the viral gene product VP1. However, the 5' and 3' non-translated regions of this gene are left and thus replacement of the VP1 sequences with foreign DNA permits efficient expression of the latter.

swing-out rotor A centrifuge rotor having buckets which are free to swing out into a horizontal position when the rotor is in motion. The long dimension of the centrifuge tube is thus held parallel to the lines of centrifugal force, and a density gradient may be formed along the entire length of the tube. This gradient will therefore cover a greater density range than in an angle or vertical rotor. Swing-out rotors are particularly useful when there is uncertainty about the buoyant densities of the molecules to be resolved.

T

tac promoter A hybrid promoter constructed *in vitro* which contains the −35 sequence of the *trp* promoter and the −10 sequence of the *lacZ* promoter. A pair of such promoters, *tac* I and *tac* II were constructed with minor variations between them; both are repressed by the *lac* repressor and induced by IPTG. *Tac* I is three times, and *tac* II twice, as efficient as a depressed *trp* promoter.

tailing The addition of a stretch of identical nucleotides to the ends of a restriction fragment using the enzyme terminal transferase. Complementary homopolymeric tails may be placed on two different molecules and then annealed to produce a recombinant. (*See* A's and T's method and dG·dC tailing.) If the tails are sufficiently long, the recombinant will be stable *in vitro* and not require ligation before transformation into a host organism.

tandem array The contiguous arrangement of two or more identical sequences within a DNA molecule.

e.g.   XYZDEFGABCABCABCABCABCJKLSTUV
tandem array

T antigen (pronounced large tee antigen) An enzyme encoded by SV40 which is essential for viral DNA replication. It is also essential for cellular transformation and was named T for transforming antigen.

target site Restriction endonucleases bind to DNA at a specific sequence known as their recognition site and then cut the DNA within a sequence known as the target site. For Type II restriction endonucleases, the kind most commonly used in genetic engineering experiments, the recognition site and the target site are one and the same thing.

TATA box, Goldberg–Hogness box, Goldbrick TATA is the canonical sequence which occurs within the region of a eukaryotic promoter which is thought to be analogous to the Pribnow box of prokaryotes. The TATA sequence facilitates, but is not necessarily essential for, transcription.

Tc^r, Tc^s Tetracycline resistant, tetracycline sensitive.

T-DNA, transferred-DNA The segment of DNA from a Ti or Ri plasmid which is transferred from *Agrobacterium* to the genome of its plant host and causes tumour formation or root induction. In many cases the

T-DNA is stably inherited by plants regenerated from Ti or Ri plasmid-induced callus tissue. Foreign DNA inserted into the T-DNA can also be stably inherited by such a regenerated plant.

TE buffer A shorthand nomenclature for a Tris-HCl buffer solution, pH 7.5, containing the chelating agent ethane diamine tetraacetic acid (EDTA). It is commonly used to dissolve nucleic acids, particularly DNA. The inclusion of EDTA is to sequester any heavy metal cations and prevent them from damaging the DNA.

telomere The end of a eukaryotic chromosome. Telomeres have a special nucleotide sequence which facilitates the replication of a linear duplex.

temperate Bacteriophages are described as temperate if they have two alternative developmental pathways. Following the infection of the host they may enter either the lytic or the lysogenic cycle.

template, template strand The nucleic acid strand which is copied during replication or transcription.

terminal transferase, deoxynucleotidyl transferase An enzyme commonly isolated from calf thymus which will add deoxynucleotide triphosphates (dNTP) to a 3'-OH group on a DNA fragment. It is used in the construction of recombinant molecules. The vector which has been opened at a single site is incubated with terminal transferase and one of the four kinds of dNTP; this is called tailing or homopolymer tailing. The DNA fragments to be inserted into the vector are tailed with the complementary dNTP, i.e. if the vector is tailed with dATP then dTTP is added to the inserts. These single-strand extensions are then annealed with one another to form hybrid molecules which can be used to transform bacteria. Any gaps or nicks in the hybrid are repaired by the bacteria and no DNA ligase is used in the procedure. (*See* A's and T's method)

termination codon, stop codon The trinucleotide sequences in mRNA molecules which terminate protein synthesis. There are three termination codons: UAA (ochre), UAG (amber) and UGA (opal). (*See* nonsense mutation, suppressor)

terminator (of transcription) A DNA sequence just downstream of the coding segment of a gene which is recognised by RNA polymerase as a signal to stop synthesising mRNA. In prokaryotes, terminators usually have an inverted repeat followed by a short stretch of Us at the very end of the transcribed portion. There may also be sequences beyond the transcribed part of the gene which influence the termination of transcription.

terminator (of translation) *See* termination codon.

tetracycline resistance Tc^r, tetracycline sensitive Tc^s Resistance or sensitivity to the lethal effects of the antibiotic tetracycline. Tetracycline is bacteriostatic, it halts growth by preventing the binding of charged tRNAs to the ribosome. Growth can resume again if the tetracycline is removed from the culture. Tetracycline-sensitive cells can be enriched in a population of tetracycline-resistant cells by a technique called cycloserine enrichment where Tc^r cells are lysed while Tc^s cells are not. Recombinant clones which have DNA fragments inserted into the tetracycline-resistance gene of a cloning vector, such as pBR322, can therefore be selectively enriched.

T4, phage T4 A complex DNA-containing bacteriophage which infects *E. coli*. It has been an experimental subject for many years and a study of mutations in one region of T4 provided the first strong evidence for the triplet nature of the genetic code. Many enzymes which are of enormous value as tools in molecular biology are encoded by T4.

T4 DNA ligase An enzyme encoded by phage T4 which catalyses the ATP-dependent formation of a phosphodiester bond between a 5′ phosphate and a 3′ hydroxyl in duplex DNA. The enzyme will efficiently join together two DNA fragments which have the same sticky ends or seal a nick within a duplex DNA molecule. At high enzyme concentrations T4 DNA ligase will join two blunt-ended DNA fragments together. It is the ligase most commonly used in recombinant DNA experiments.

T4 DNA polymerase An enzyme encoded by phage T4 which catalyses the synthesis of DNA in the 5′ to 3′ direction using a template and a primer with a 3′ hydroxyl group. The enzyme has a 3′ to 5′ exonuclease activity but, unlike *E. coli* DNA polymerase I, T4 DNA polymerase does not have a 5′ to 3′ exonuclease activity. Therefore it cannot be used for labelling DNA by nick translation but can be used to label DNA to high specific activity in the following manner: the DNA is first incubated with T4 DNA polymerase in the absence of dNTPs and is consequently degraded by the enzyme's 3′ to 5′ exonuclease activity. The four dNTPs are then added to the reaction mix which permits the resynthesis of the degraded strand by the enzyme's 5′ to 3′ polymerase activity. If the α-phosphate of one of the four dNTPs is labelled with ^{32}P a highly radioactive product will result.

T4 polynucleotide kinase An enzyme encoded by phage T4 which catalyses the transfer of a phosphate moiety from ATP to the 5′ hydroxyl group of a polynucleotide. One of its main uses is to label

DNA and RNA by the transfer of a ^{32}P-phosphate group from γ^{32}P-ATP. DNA labelled in this way can then be sequenced by the Maxam and Gilbert technique.

T4 RNA ligase An enzyme encoded by phage T4 which will join DNA ends to DNA, RNA ends to RNA, and DNA ends to RNA provided one end has a 5′ phosphate and the other has a 3′ hydroxyl. The enzyme requires ATP. It is used to form recombinant RNA and will greatly increase the rate of blunt-end joining of two DNA fragments with T4 DNA ligase.

3′, three prime The atoms making up the ring of a purine or pyrimidine base are numbered 1, 2, 3 etc. In order to distinguish them, the constituent atoms of the pyranose ring of the ribose or deoxyribose sugar to which the base is attached in a nucleoside are numbered 1′, 2′, 3′ etc. It is the 3′ carbon to which the phosphate group of the next nucleotide in the chain is attached. (*See* 5′ for the ring-numbering system)

3′ extension A single-stranded tail left on the 3′ end of a DNA molecule following digestion with a restriction endonuclease. E.g. the enzyme *Pst* I forms a four base pair 3′ extension:

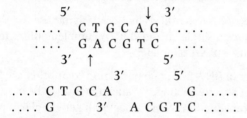

(*See* sticky ends)

Ti plasmid A class of large conjugative plasmids found in the soil bacterium *Agrobacterium tumefaciens* and responsible for the crown gall disease of broad-leaved plants. A segment of the Ti plasmid (the T-DNA) is found in the genome of tumour tissue of plants with crown gall disease. Ti plasmids and particularly the T-DNA have generated considerable interest as they may be useful for the insertion of foreign genes into plants.

tk⁻, thymidine kinase deficient tk⁻ mutants are unable to phosphorylate the DNA precursor, thymidine. tk⁻ cell lines are often used in mammalian somatic cell genetics since they are resistant to the thymidine analogue, bromodeoxyuridine (BUdR).

T_m *See* melting temperature.

Tn3 A bacterial transposon which carries the gene for ampicillin resistance as well as genes for its own transposition. Tn3 was the first transposon to have its DNA sequence completely determined. The ampicillin-resistance gene from Tn3 was used in the construction of pBR322. Tn3 and the closely related transposon Tn1 are also known as TnA.

Tn5 A bacterial transposon which carries a gene encoding resistance to the antibiotics kanamycin and neomycin, as well as genes involved in transposition. It is a well characterised transposon, the complete sequence of which is known. The antibiotic resistance gene has been used in eukaryotic cloning vectors to give resistance to the aminoglycoside antibiotic, G418.

Tn10 A bacterial transposon which carries the genes for tetracycline resistance as well as genes involved in its own transposition.

tra The genetic notation for the genes involved in the transfer of a plasmid from one cell to another.

transcript The RNA product of a gene. The primary transcript produced by RNA polymerase must often be processed or modified in order to form the mature, functional mRNA, rRNA or tRNA species.

transcription The process of RNA synthesis by RNA polymerase (transcriptase) to produce a single-stranded RNA complementary to a DNA or, in certain viruses, RNA template.

transduction The transfer of DNA from one bacterium to another via the release and infection of a bacteriophage. Transduction may be of two kinds: (i) Specialised transduction where a bacterial gene replaces part of the viral genome. Bacterial genes transferred are those which flank the integration site of a temperate bacteriophage such as λ. (ii) Generalised transduction, where only bacterial DNA is packaged into the phage particle, completely replacing the phage genome. This arises from the random destruction of the host bacterium's genome during lytic infection with the phage, followed by the packaging of an appropriately sized fragment of bacterial DNA into the phage particle. In both cases the transducing phage is defective and cannot establish a lytic infection in the new host.

transfection The transformation of bacteria with phage DNA.

transferable A term used to describe plasmids which are able to promote their own transfer from bacterium to bacterium.

transformation A mechanism of gene transfer which involves the uptake of purified DNA by microorganisms. Following entry into the cell, the transforming DNA may recombine with that of the host or may replicate independently as a plasmid or temperate phage. Cells able to take up DNA are described as competent. Some organisms, e.g. *Bacillus* and *Diplococcus* species, are naturally competent. Others, e.g. *E. coli* and yeast, must be subjected to various treatments which induce the transfer of DNA across the cell membrane. Transformation with phage DNA is often described as transfection. Transformation is an essential step in most genetic engineering experiments.

Transformation is also used in mammalian cell genetics to describe the conversion of a cell from the normal to the tumorous state.

translation The process of protein synthesis carried out by ribosomes which de-code the information contained in mRNA.

transmissible *See* transferable.

transplacement, gene transplacement, replacement, gene replacement A technique, originally developed in yeast molecular biology, for exchanging a DNA sequence on a recombinant plasmid for the equivalent sequence in one of the host's chromosomes. Thus a wild-type copy of a gene can be replaced by a mutant copy or *vice versa*. The method may have implications for gene therapy in domestic animals and Man.

The technique operates as shown in the diagram (Fig. 33). The host organism is transformed with an integrative plasmid (e.g. *see* YIp). The recombinant plasmid integrates itself into the host's chromosome by homologous recombination, thereby creating a duplication of the sequences of interest. A subsequent intrachromosomal recombination event loops out the integrated plasmid together with the copy (x) of the gene originally contained in the chromosome. The copy (X) of the gene originally contained in the recombinant plasmid is left behind and is said to have transplaced (x). (Fig. 33)

transposition The process whereby a transposon or insertion sequence inserts into a new site on the same or another DNA molecule. The exact mechanism is not fully understood and different transposons may transpose by different mechanisms. Transposition in bacteria does not require extensive DNA homology between the transposon and the target DNA. The phenomenon is therefore described as illegitimate recombination.

transposon A discrete piece of DNA which can insert itself into many different sites in other DNA sequences within the same cell. The proteins necessary for the transposition process are encoded within the transposon. A copy of the transposon is retained at the original site after transposition; thus DNA replication is involved. The ends of a transposon are usually identical but in inverse orientation with respect to one another (an inverted repeat). The insertion of a transposon into a new site generates a direct repeat of the target DNA of 3 to 11 base pairs. The amount of host DNA which is repeated as a consequence of transposition is specific for each transposon. Transposons in bacteria often carry genes which confer antibiotic or heavy-metal resistance upon the host cell and are designated Tn1, 2, 3 etc. Transposons have been identified in many organisms including bacteria, yeast, *Drosophila* and humans.

Fig. 33. Transplacement.

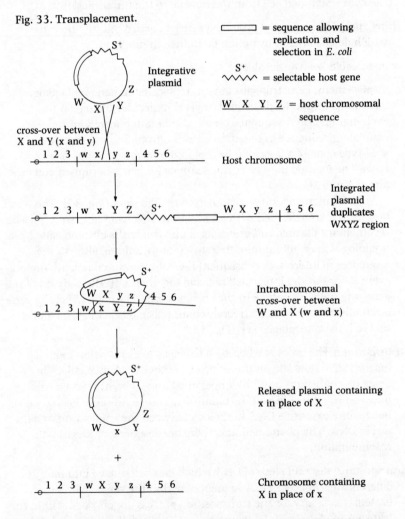

trimethoprimʳ, trimethoprimˢ Resistance or sensitivity to the inhibitory effects of the antibacterial compound trimethoprim. This drug is a

competitive inhibitor of the enzyme dihydrofolate reductase (DHFR)
which is involved in generating nucleosides for DNA synthesis.
Bacteria resistant to trimethoprim have often acquired a plasmid gene
which codes for a dihydrofolate reductase which is almost insensitive
to trimethoprim. These bacterial genes also confer methotrexate
resistance to eukaryotic cells, which are insensitive to trimethoprim.
With the correct transcription signals, therefore, the bacterial
trimethoprim-resistance gene can be used as a selectable marker for
cloning vectors in animal and plant cells. One such gene is found on a
transposon, Tn7.

triplet A sequence of three nucleotides which specifies an amino acid.
The elucidation of the genetic code involved the binding of charged
tRNA species to chemically synthesised ribonucleotide triplets. (A full
list of codon assignments can be found in Appendix 5)

Tris-acetate buffer A buffer consisting of Tris base and acetic acid. It is
commonly used in the electrophoresis of DNA or RNA through agarose
or polyacrylamide gels. It contains 40 mM Tris base, 20 mM Na
acetate, and 1 mM EDTA adjusted to pH 8.0 with glacial acetic acid.

Tris base An organic base which is often used to make buffer solutions
for dissolving nucleic acids. Its full name is
tris(hydroxymethyl)aminomethane.

Tris-borate buffer A buffer consisting of Tris base and boric acid. It is one
of the common buffers for the electrophoresis of DNA and RNA
through agarose and polyacrylamide gels. It consists of 89 mM Tris
base, 89 mM boric acid, and 2 mM EDTA (pH 8.0).

tRNA, transfer RNA The RNA molecule which is responsible for decoding
the genetic information contained within mRNA. Transfer RNA
molecules are folded into a 'clover-leaf' secondary structure by
intrastrand base pairing as shown in Fig. 34.
    The anticodon loop contains a nucleotide triplet complementary to a
specific codon within the mRNA molecule. Each tRNA is 'charged'
with the correct amino acid molecule, via its 3' adenosine moiety, by
an enzyme called aminoacyl-tRNA synthetase.

***trp* promoter** (pronounced trip) The promoter of the *E. coli* tryptophan
synthesis operon. It is a strong promoter which is inducible by
incubating cells with 3-β-indoleacrylic acid. The −35 sequence of the
trp promoter was used in constructing the *tac* promoters.

2μ circle A circular DNA plasmid found in the nuclei of the yeast,
Saccharomyces cerevisiae. The molecule is 2 μm long and contains 6318
bp including two 599 bp repeat sequences. Each haploid nucleus

contains about 50 copies of the 2μ circle. The molecule has no readily detectable phenotype, but contains four genes which are involved in its own replication, amplification and recombination. The replication functions of the 2μ molecule are exploited in the YEp (yeast episomal plasmids) series of vectors, used in yeast cloning experiments.

Fig. 34. A diagram of all tRNA sequences except for initiator tRNAs. The position of invariant and semi-invariant bases is shown. The numbering system is that of yeast tRNA^{Phe}. Y stands for pyrimidine, R for purine, H for a hypermodified purine. (From Rich & RajBhandary (1976), *Annual Review of Biochemistry*, **45**, 805–60.)

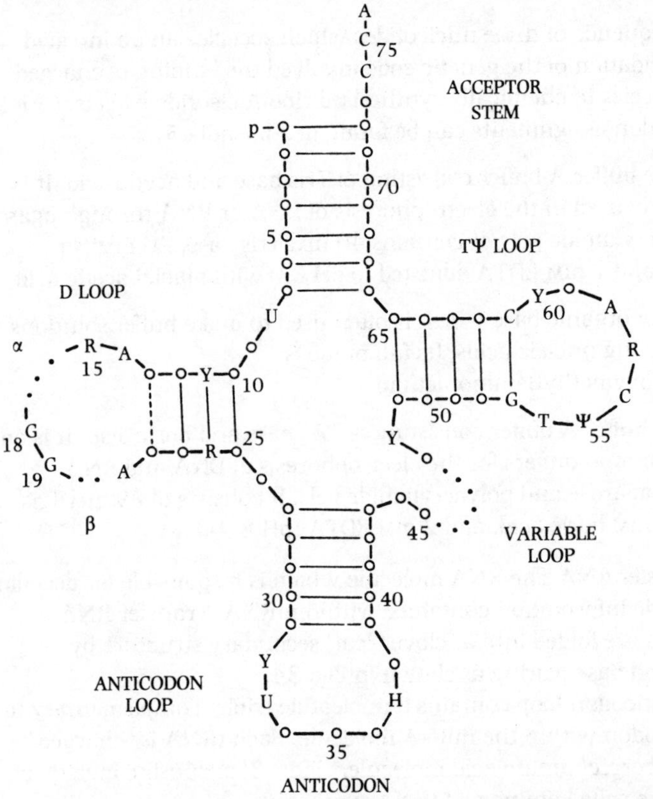
ACCEPTOR
STEM
TΨ LOOP
D LOOP
VARIABLE
LOOP
ANTICODON
LOOP
ANTICODON
75
70
5
60
65
15
10
50
55
18
25
19
45
30
40
35
α
β

two-step ligation A gene cloning protocol in which linearised vector molecules and insert DNA fragments are mixed and ligated at high DNA concentration. This favours the joining of vector and insert DNA. The DNA solution is then diluted for the second step of the ligation. This favours the circularization of recombinant molecules.

Ty1 A yeast transposon. A transposable genetic element found in most strains of *Saccharomyces cerevisiae*. Ty1 is *ca.* 5.6 kb long and has two terminal direct repeat sequences, known as δ, of 250 bp. Ty1 preferentially inserts into promoter regions of genes causing the duplication of 5 bp of its target sequence. It may alter the expression of the gene into which it inserts, often placing it under mating-type control.

U

UAS, upstream activator sequence A part of a eukaryotic promoter some distance away from both the transcriptional start site and the TATA box. The UAS affects the extent of transcription from a given promoter without affecting the site of transcriptional initiation. UASs may be involved in the correct positioning or phasing of nucleosomes to facilitate RNA polymerase binding further downstream. Alternatively they may represent the point at which the RNA polymerase enters the chromatin complex.

ultracentrifuge An instrument which can spin a rotor at great speed, 65 000–100 000 revolutions per minute, to create forces of up to 420 000 times that of gravity. These centrifugal forces are used to separate cells, particles or molecules according to either their size or density. There are two types of ultracentrifuge. The preparative ultracentrifuge is used to prepare purified samples of biological material for further experimentation or analysis. An analytical ultracentrifuge is used to measure directly the position of the sample in the rotor chamber while the rotor is in motion. To this end, the walls of the centrifuge tube are made of quartz to enable the behaviour of the sample to be observed by either UV or Schlieren optics. (*See* density gradient centrifugation, sucrose gradient, CsCl gradient)

U1–U6 RNA, U snRNAs A family of stable, small nuclear RNA molecules originally found in rat Novikoff hepatoma cells but subsequently discovered in a range of mammals and other higher eukaryotes. The U snRNAs range in size from 90 to 400 nucleotides. U1 RNA is present in *ca.* 10^6 copies per cell while the others have a copy number of 10^4–10^5. All the U snRNAs have a 5′ cap structure analogous to that of mRNA and, unlike other small RNAs, are synthesised by RNA polymerase II. They are found in ribonucleoprotein complexes called snurps.

U1 RNA is thought to act as an external guide sequence to facilitate the correct splicing of introns from mRNA precursor molecules. It contains sequences at its 5′ end which are complementary to the conserved intron sequences near the intron/exon junction (splice junction).

The function of the other U snRNAs is less clear. It has been suggested that U3 RNA is involved in the processing of pre-rRNA, a role consistent with its localisation within the nucleolus. U4 RNA may facilitate the polyadenylation of mRNA by hybridising to conserved sequences near the 3' end of mRNA precursor molecules.

up-promoter mutant A strain carrying a mutation in a promoter sequence which increases transcription from that promoter.

upstream In the opposite direction to that in which a nucleic acid or protein molecule is synthesised i.e. on the 5' side of any given site in DNA or RNA and on the N-terminal side of any site within a polypeptide.

When referring to upstream regions of a gene with respect to its activity as a template for RNA synthesis, it is the 3' side of any given region in the coding (or sense) strand which is denoted.

URF, unidentified reading-frame An open reading-frame recognised from a DNA sequence for which no genetic function is known.

V

vector A general term applied to a DNA molecule, derived from a plasmid or bacteriophage, into which fragments of DNA may be inserted or cloned. The vector should contain one or more unique restriction sites for this purpose and be capable of autonomous replication in a defined host or vehicle organism such that the cloned sequence is reproduced. The vector molecule should confer some well defined phenotype on the host organism which is either selectable, e.g. drug resistance, or readily detected, e.g. plaque formation. (*See* shuttle vector, bifunctional vector, YCp, YEp, YIp, YRp and a range of specific vectors under p for plasmid)

vehicle The host organism used for the replication or expression of a cloned gene or other sequence. (Compare with vector, the DNA molecule which contains the cloned gene.) The term is little used and is often confused with vector.

vertical rotor A centrifuge rotor in which the wells holding the tubes are drilled parallel to the axis of rotation and at right angles to the lines of centrifugal force. A density gradient, therefore, is formed only across the width of the tube. This means that gradients are rapidly formed and run-times are significantly reduced as compared with angle and, especially, swing-out rotors. The reorientation of the gradient as the rotor comes to rest may cause mixing problems however.

virulent phage A bacteriophage whose life cycle always involves the lysis of its host. (*cf.* lysogenic phage)

W

Watson–Crick base-pairing The pattern of complementary base-pairing, G with C and A with T, found in the B-form of DNA. It is named after James Watson and Francis Crick who first proposed the double helix model of DNA structure.

Western blot A procedure analogous to Southern transfer but in this case proteins are transferred from a polyacrylamide gel onto a suitable immobilising matrix, e.g. a nitrocellulose sheet. The proteins attached to the support matrix can then be probed with, for example, a specific antibody to identify a particular protein species. The transfer from the gel to the matrix is often carried out by electro-blotting.

wheat-germ system *See in vitro* translation.

wild-type The usual or non-mutant form of a gene or organism. This term was originally meant to denote the form in which the organism was usually found in Nature (the 'wild'). It has come to have a more specialised meaning. Wild-type refers to the genetic constitution of an organism at the start of a programme of mutagenesis. Thus a geneticist's wild-type strain may already contain a number of mutations (markers) before he sets about introducing further changes.

X

X-gal A shorthand nomenclature for the compound 5-chloro-4-bromo-3-indolyl-β-D-galactoside. It is a colourless compound which, when hydrolysed by a β-galactosidase, releases a bright blue non-diffusible indigo dye. It is used as a colour indicator for bacteria able to use lactose and in the M13mp phage cloning system.

***Xma* I** (sometimes pronounced exmar one) A type II restriction enzyme from the bacterium *Xanthomonas malvacearum* which recognises the DNA sequence shown below and cuts it at the sites indicated by the arrows:

$$\downarrow$$
$$5'\ \ C\ C\ C\ G\ G\ G\ \ 3'$$
$$3'\ \ G\ G\ G\ C\ C\ C\ \ 5'$$
$$\uparrow$$

The *Xma* I enzyme is a isoschizomer of *Sma* I. (A full list of restriction enzymes can be found in Appendix 1)

Y

YCp, Yeast centromeric plasmid A *S. cerevisiae* cloning vector which contains a cloned centromere sequence. The centromere confers stability on the plasmid but also reduces its copy number to one or two per cell. At meiosis, a YCp segregates as a minichromosome.

In addition to the centromeric sequence, a YCp must contain a chromosomal replicator (*ARS* sequence) and a selectable yeast gene. Frequently an *E. coli* origin is also included to permit replication in that organism. pYe(CEN3)41 is a typical YCp, it contains the centromere from yeast chromosome III, the *ARS*1 replicator which is found adjacent to the yeast *TRP*1 gene, *LEU*2 as a selectable gene and pBR322 sequences to permit replication in *E. coli*, as shown below.

Fig. 35. YCp (map).

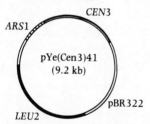

yeast Unless specified this is taken to mean *Saccharomyces cerevisiae*. However gene-cloning systems have been developed for the following unicellular fungi:

Saccharomyces cerevisiae, the budding yeast used in the production of bread and ale.

Schizosaccharomyces pombe, the fission yeast used to produce beer in some parts of Africa.

Kluyveromyces lactis, the lactose-fermenting yeast used in the processing of whey.

Yarrowia (Saccharomycopsis) lipolytica, the hydrocarbon-utilising yeast used in the production of organic acids.

YEp, Yeast episomal plasmid A *S. cerevisiae* cloning vector which relies, for its replication, on the 2μ plasmid origin. YEps also contain a selectable yeast gene and, often, an *E. coli* origin. pJDB219 is a typical YEp.

YEps transform yeast at high frequency (10^3–10^4 transformants per μg of DNA) and can be quite stable.

Fig. 36. YEp (map of pJDB219)

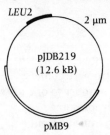

YIp, Yeast integrative plasmid A *S. cerevisiae* cloning vector which
consists of a selectable yeast gene inserted into an *E. coli* plasmid, e.g.
pBR322. A YIp does not contain a yeast replicator and thus can only
be maintained if integrated into a chromosome by, usually,
homologous recombination. CV9 is a typical YIp (see below).

Fig. 37. YIp (map of CV9).

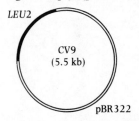

It contains the yeast *LEU2* gene cloned into pBR322. It will integrate
into the non-functional *leu2* gene on chromosome III of the recipient
cell.

YIps transform yeast with low efficiency (10–100 transformants per
μg of DNA) but are very stable.

YRp, Yeast replicative plasmid A *S. cerevisiae* cloning vector which uses a
chromosomal replicator (*ARS* sequence) to replicate in yeast. YRps will
also contain a selectable yeast gene and an *E. coli* origin. YRps
transform yeast at high frequency (*ca.* 10^3 transformants per μg of
DNA) but are very unstable even under selective conditions. YRp 17 is
a typical YRp (see below).

Fig. 38. YRp (map of YRp17).

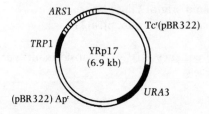

Z

Z-DNA, zig-zag DNA A form of DNA duplex in which the double helix is wound in a left-hand, instead of a right-hand, manner. DNA adopts the Z configuration when purines and pyrimidines alternate on a single strand, e.g.

    CGCGCGCG   or   CACACACACA
    GCGCGCGC   GTGTGTGTGT

Z-DNA is known to exist within eukaryotic chromosomes but its function is presently obscure.

List of restriction enzymes

This list of the restriction enzymes so far isolated is adapted from the publication by Roberts (1983) *Nucleic Acids Research* 11, r135–r167. The sequence of recognition sites is written 5′ to 3′ and only one strand is given. The point where the enzyme cleaves the nucleotide strand is indicated, when known, by an arrow. Bases known to be modified by a specific methylase are indicated by an asterisk. Where two or more enzymes have the same recognition site (i.e. are isoschizomers) then the code for the first enzyme to be found to have that site is given in parentheses.

Microorganism	Enzyme	Sequence	λ	pBR322
				Number of cleavage sites
Acetobacter aceti	*Aat*I (*Stu*I)	AGGCCT	5	0
	*Aat*II	GACGT↑C	7	1
Acetobacter aceti sub. *liquefaciens*	*Aac*I (*Bam*HI)	GGATCC	5	1
Acetobacter aceti sub. *liquefaciens*	*Aae*I (*Bam*HI)	GGATCC	5	1
Acetobacter aceti sub. *Orleanensis*	*Aor*I (*Eco*RII)	CC↑(A/T)GG	>35	6
Acetobacter pasteurianus sub. *pasteurianus*	*Apa*I	GGGCC↑C	1	0
Achromobacter immobilis	*Aim*I	?	?	?
Acinetobacter calcoaceticus	*Acc*I	GT↑(A/C) (G/T)AC	7	2
	*Acc*II (*Fnu*DII)	CGCG	>50	23
	*Acc*III	?	>10	?
Actinomadura madurae	*Ama*I (*Nru*I)	TCGCGA	7	1
Agmenellum quadruplicatum	*Aqu*I (*Ava*I)	CPyCGPuG	8	1
Agrobacterium tumefaciens	*Atu*AI	?	>30	?
Agrobacterium tumefaciens B6806	*Atu*BI (*Eco*RII)	CC(A/T)GG	>35	6
Agrobacterium tumefaciens LIBV7	*Atu*BVI	?	>14	?
Agrobacterium tumefaciens ID 135	*Atu*II (*Eco*RII)	CC(A/T)GG	>35	6
Agrobacterium tumefaciens C58	*Atu*CI (*Bcl*I)	TGATCA	7	0
Alcaligenes species	*Asp*AI (*Bst*EII)	G↑GTNACC	11	0
Anabaena catanula	*Aca*I	?	?	?
Anabaena cylindrica	*Acy*I	GPu↑CGPyC	>14	6
Anabaena flos-aquae	*Afl*I (*Ava*II)	G↑G(A/T)CC	>17	8
	*Afl*II	C↑TTAAG	3	0
	*Afl*III	A↑CPuPyGT	>30	1
Anabaena oscillarioides	*Aos*I (*Mst*I)	TGC↑GCA	>10	4
	*Aos*II (*Acy*I)	GPu↑CGPyC	>14	6
Anabaena strain Waterbury	*Ast*WI (*Acy*I)	GPu↑CGPyC	>14	6
Anabaena subcylindrica	*Asu*I	G↑GNCC	>30	15
	*Asu*II	TT↑CGAA	7	0
	*Asu*III (*Acy*I)	GPu↑CGPyC	>14	6
Anabaena variabilis	*Ava*I	C↑PyCGPuG	8	1
	*Ava*II	G↑G(A/T)CC	>17	8
	*Ava*III	ATGCAT	?	0
Anabaena variabilis uw	*Avr*I (*Ava*I)	CPyCGPuG	8	1
	*Avr*II	CCTAGG	2	0
Aphanothece halophytica	*Aha*I (*Cau*II)	CC(C/G)GG	>30	10
	*Aha*II	?	?	?

Microorganism	Enzyme	Sequence	Number of cleavage sites	
			λ	pBR 322
	AhaIII	TTT↑AAA	13	3
Arthrobacter luteus	AluI	AG↑CT	> 50	16
Arthrobacter pyridinolis	ApyI (EcoRII)	CC↑(A/T)GG	> 35	6
Bacillus acidocaldarius	BacI (SacII)	CCGCGG	4	0
Bacillus amyloliquefaciens F	BamFI (BamHI)	GGATCC	5	1
Bacillus amyloliquefaciens H	BamHI	G↑GATCC	5	1
Bacillus amyloliquefaciens K	BamKI (BamHI)	GGATCC	5	1
Bacillus amyloliquefaciens N	BamNI (BamHI)	GGATCC	5	1
	BamN$_x$ (AvaII)	G↑G(A/T)CC	> 17	8
Bacillus aneurinolyticus	BanI (HgiCI)	GGPyPuCC	13	9
	BanII (HgiJII)	GPuGCPy↑C	7	2
	BanIII (ClaI)	ATCGAT	15	1
Bacillus brevis S	BbvSI	GC(*A/T)GC	specific methylase	
Bacillus brevis	BbvI	GCAGC (8/12)	↑30	21
Bacillus caldolyticus	BclI	T↑GATCA	7	0
Bacillus centrosporus	BcnI (CauII)	CC(C/G)↑GG	> 50	10
Bacillus cereus	Bce14579	?	> 10	?
Bacillus cereus	Bce1229	?	> 10	?
Bacillus cereus	Bce170 (PstI)	CTGCAG	18	1
Bacillus cereus Rf sm st	BceR (FnuDII)	CGCG	> 50	23
Bacillus globigii	BglI	GCCNNNN↑NGGC	22	3
	BglII	A↑GATCT	6	0
Bacillus megaterium 899	Bme899	?	> 5	?
Bacillus megaterium B205-3	Bme205	?	> 10	?
Bacillus megaterium	BmeI	?	> 10	?
Bacillus pumilus AHU1387A	BpuI	?	6	?
Bacillus sphaericus	Bsp1286	?	> 30	?
Bacillus sphaericus R	BspRI (HaeIII)	GG↑CC	> 50	22
Bacillus stearothermophilus C1	BstCI (HaeIII)	GGCC	> 50	22
Bacillus stearothermophilus C11	BssCI (HaeIII)	GGCC	> 50	22
Bacillus stearothermophilus G3	BstGI (BclI)	TGATCA	7	0
	BstGII (EcoRII)	CC(A/T)GG	> 35	6
Bacillus stearothermophilus G6	BssGI (BstXI)	GGANNNNNNTCC	10	0
	BssGII (MboI)	GATC	> 50	22
Bacillus stearothermophilus H1	BstHI (XhoI)	CTCGAG	1	0
Bacillus stearothermophilus H3	BssHI (XhoI)	CTCGAG	1	0
	BssHII (BsePI)	GCGCGC	6	0
Bacillus stearothermophilus H4	BsrHI (BsePI)	GCGCGC	6	0
Bacillus stearothermophilus P1	BssPI	?	> 30	?
Bacillus stearothermophilus P5	BsrPI	?	11	0
	BsrPII (MboI)	GATC	> 50	22
Bacillus stearothermophilus P6	BsePI	GCGCGC	6	0
Bacillus stearothermophilus P8	BsaPI (MboI)	GATC	> 50	22
Bacillus stearothermophilus P9	BsoPI (BsrPI)	?	11	0
Bacillus stearothermophilus T12	BstTI (BstXI)	GGANNNNNNTCC	10	0
Bacillus stearothermophilus XI	BstXI	GGANNNN↑NTCC	10	0
	BstXII (MboI)	GATC	> 50	22
Bacillus stearothermophilus 1503–4R	BstI (BamHI)	G↑GATCC	5	1
Bacillus stearothermophilus 240	BstAI	?	?	?
Bacillus stearothermophilus ET	BstEI	?	?	?
	BstEII	G↑GTNAAC	11	0
	BstEIII (MboI)	GATC	> 50	22
Bacillus stearothermophilus	BstPI (BstEII)	G↑GTNACC	11	0
Bacillus stearothermophilus	BstNI (EcoRII)	CC↑(A/T)GG	> 35	6
Bacillus stearothermophilus strain 822	BseI (HaeIII)	GGCC	> 50	22
	BseII (HpaI)	GTTAAC	13	0

127

Microorganism	Enzyme	Sequence	Number of cleavage sites λ	pBR322
Bacillus subtilis strain R	*Bsu*RI (*Hae*III)	GG↑CC	>50	22
Bacillus subtilis Marburg 168	*Bsu*M	?	>10	?
Bacillus subtilis	*Bsu*6633	?	>20	?
Bacillus subtilis	*Bsu*1076 (*Hae*III)	GGCC	>50	22
Bacillus subtilis	*Bsu*1114 (*Hae*III)	GGCC	>50	22
Bacillus subtilis	*Bsu*1247 (*Pst*I)	CTGCAG	18	1
Bacillus subtilis	*Bsu*1145	?	>20	?
Bacillus subtilis	*Bsu*1192I (*Hpa*II)	CCGG	>50	26
	*Bsu*1192II (*Fnu*DII)	CGCG	>50	23
Bacillus subtilis	*Bsu*1193 (*Fnu*DII)	CGCG	>50	23
Bacillus subtilis	*Bsu*1231I (*Hpa*II)	CCGG	>50	26
	*Bsu*1231II (*Fnu*DII)	CGCG	>50	23
Bacillus subtilis	*Bsu*1259	?	>8	?
Bifidobacterium bifidum YIT4007	*Bbi*I (*Pst*I)	CTGCAG	18	1
	*Bbi*II (*Acy*I)	GPuCGPyC	>14	6
	*Bbi*III (*Xho*I)	CTCGAG	1	0
	*Bbi*IV	?	?	0
Bifidobacterium breve	*Bde*I(*Nar*I)	GGCGC↑C	>2	4
Bifidobacterium breve S1	?	?	?	?
Bifidobacterium breve S50	*Bbe*AI (*Nar*I)	GGCGCC	>2	4
	*Bbe*AII	?	?	?
Bifidobacterium infantis 659	*Bin*I	GGATC	>10	6
Bifidobacterium infantis S76e	*Bin*SI (*Eco*RII)	CC(^A_T)GG	>35	6
	*Bin*SII (*Nar*I)	GGCGCC	>2	4
Bifidobacterium longum E194b	*Blo*I	?	?	?
Bifidobacterium thermophilum RU326	*Bth*I (*Xho*I)	CTCGAG	1	0
	*Bth*II	?	?	0
Bordetella bronchiseptica	*Bbr*I (*Hin*dIII)	AAGCTT	6	1
Bordetella pertussis	*Bpe*I (*Hin*dIII)	AAGCTT	6	1
Brevibacterium albidum	*Bal*I	TGG↑CCA	15	1
Brevibacterium luteum	*Blu*I (*Xho*I)	C↑TCGAG	1	0
	*Blu*II (*Hae*III)	GGCC	>50	22
Calothrix scopulorum	*Csc*I (*Sac*II)	CCGC↑GG	4	0
Caryophanon latum L	*Cla*I	AT↑CGAT	15	1
Caryophanon latum	*Clm*I (*Hae*III)	GGCC	>50	22
	*Clm*II (*Ava*II)	GG(^A_T)CC	>17	8
Caryophanon latum	*Clt*I (*Hae*III)	GG↑CC	>50	22
Caryophanon latum RII	*Clu*I	?	>20	?
Caryophanon latum H7	*Cal*I	?	14	?
Caulobacter crescentus CB-13	*Ccr*I	?	1	1
	*Ccr*II (*Xho*I)	CTCGAG	1	0
Caulobacter fusiformis	*Cfu*I (*Dpn*I)	GATC	only cleaves methylated DNA	
Chloroflexus aurantiacus	*Cau*I (*Ava*II)	GG(^A_T)CC	>30	
	*Cau*II	CC↑(^G_C)GG	>50	
Chromatium vinosum	*Cyn*I (*Sau*I)	CC↑TNAGG	>10	
Chromobacterium violaceum	*Cvi*I	?	?	
Citrobacter freundii	*Cfr*I	Py↑GGCCPu	>25	6
Clostridium formicoaceticum	*Cfo*I (*Hha*I)	GCGC	>50	31
Clostridium pasteurianum	*Cpa*I (*Mbo*I)	GATC	>50	22
Corynebacterium humiferum	*Chu*I (*Hin*dIII)	AAGCTT	6	1
	*Chu*II (*Hin*dII)	GTPyPuAC	34	2
Corynebacterium petrophilum	*Cpe*I (*Bcl*I)	TGATCA	7	0
Cystobacter velatus Plv9	*Cve*I	?	?	?
Desulfovibrio desulfuricans Norway strain	*Dde*I	C↑TNAG	>50	8
	*Dde*II (*Xho*I)	CTCGAG	1	0
Desulfovibrio desulfuricans	*Dds*I (*Bam*HI)	GGATCC	5	1

128

Microorganism	Enzyme	Sequence	λ	pBR322
			— Number of cleavage sites	

Microorganism	Enzyme	Sequence	λ	pBR322
Diplococcus pneumoniae	DpnI	GÁ↑TC	only cleaves methylated DNA	
Diplococcus pneumoniae	DpnII (*Mbo*I)	GATC	>50	22
Enterobacter aerogenes	EaeI (*Cfr*I)	Py↑GGCCPu	>25	6
Enterobacter cloacae	EclI	?	14	?
	EclII (*Eco*RII)	CC(A/T)GG	>35	6
Enterobacter cloacae	EcaI (*Bst*EII)	G↑GTNACC	11	0
	EcaII (*Eco*RII)	CC(A/T)GG	>35	6
Enterobacter cloacae	EccI (*Sac*II)	CCGCGG	4	0
Escherichia coli pDX1	EcoDXI	ATCA(N)₇ATTC	?	0
Escherichia coli J62 pLG74	EcoRV	GATAT↑C	14	1
Escherichia coli RY13	EcoRI	G↑AÁTTC	5	1
	EcoRI'	PuPuA↑TPyPy	>10	15
Escherichia coli R245	EcoRII	↑CC(A/T)GG	>35	6
Escherichia coli B	EcoB	TGA(N)₈TGCT		0
Escherichia coli K	EcoK	AAC(N)₆GTGC		2
Escherichia coli (PI)	EcoPI	AGÁCC		4
Escherichia coli P15	EcoP15	CAGCAG		7
Flavobacterium okeanokoites	FokI	GGATG	>50	6
Fremyella diplosiphon	FdiI (*Ava*II)	G↑G(A/T)CC	>17	8
	FdiII (*Mst*I)	TGC↑GCA	>10	4
Fusobacterium nucleatum A	FnuAI (*Hinf*I)	G↑ANTC	>50	10
	FnuAII (*Mbo*I)	GATC	>50	22
Fusobacterium nucleatum C	FnuCI (*Mbo*I)	↑GATC	>50	22
Fusobacterium nucleatum D	FnuDI (*Hae*III)	GG↑CC	>50	22
	FnuDII	CG↑CG	>50	23
	FnuDIII (*Hha*I)	GCG↑C	>50	31
Fusobacterium nucleatum E	FnuEI (*Mbo*I)	↑GATC	>50	22
Fusobacterium nucleatum 48	Fnu48I	?	>50	?
Fusobacterium nucleatum 4H	Fnu4HI	GC↑NGC	>50	42
Gluconobacter dioxyacetonicus	GdiI (*Stu*I)	AGG↑CCT	5	0
	GdiII	Py↑GGCCG	>10	5
Gluconobacter dioxyacetonicus	GdoI (*Bam*HI)	GGATCC	5	1
Gluconobacter oxydans sub. *melonogenes*	GoxI (*Bam*HI)	GGATCC	5	1
Haemophilus aegyptius	HaeI	(A/T)GG↑CC(A/T)	?	7
	HaeII	PuGCGC↑Py	>30	11
	HaeIII	GG↑CC	>50	22
Haemophilus aphrophilus	HapI	?	>30	?
	HapII (*Hpa*II)	C↑CGG	>50	26
Haemophilus gallinarum	HgaI	GACGC	>50	11
Haemophilus haemoglobinophilus	HhgI (*Hae*III)	GGCC	>50	22
Haemophilus haemolyticus	HhaI	GCG↑C	>50	31
	HhaII (*Hinf*I)	GANTC	>50	10
Haemophilus influenzae GU	HinGUI (*Hha*I)	GCGC	>50	31
	HinGUII (*Fok*I)	GGATG	>50	6
Haemophilus influenzae 173	Hin173 (*Hind*III)	AAGCTT	6	1
Haemophilus influenzae 1056	Hin1056I (*Fnu*DII)	CGCG	>50	22
	Hin1056II	?	>30	?
Haemophilus influenzae serotype b, 1076	HinbIII (*Hind*III)	AAGCTT	6	1
Haemophilus influenzae serotype c, 1160	HincII (*Hind*II)	GTPyPuAC	34	2
Haemophilus influenzae serotype c, 1161	HincII (*Hind*II)	GTPyPuAC	34	2
Haemophilus influenzae serotype e	Hin eI (*Hinf*III)	CGAAT	?	1
Haemophilus influenzae Rb	HinbIII (*Hind*III)	AAGCTT	6	1
Haemophilus influenzae Rc	HincII (*Hind*II)	GTPyPuAC	34	2
Haemophilus influenzae Rd	HindI	CÁC	specific methylase	
	HindII	GTPy↑PuÁC	34	2

129

Appendix 1 (*cont.*)

Microorganism	Enzyme	Sequence	Number of cleavage sites λ	pBR322
	HindIII	A↑AGCTT	6	1
	HindIV	GÅC	specific methylase	
Haemophilus influenzae Rf	HinfI	G↑ANTC	>50	10
	HinfII (HindIII)	AAGCTT	6	1
	HinfIII	CGAAT	?	1
Haemophilus influenzae H-1	HinHI (HaeII)	PuGCGCPy	>30	11
Haemophilus influenzae P₁	HinP₁I (HhaI)	G↑CGC	>50	31
Haemophilus influenzae S₁	HinS₁ (HhaI)	GCGC	>50	31
Haemophilus influenzae S₂	HinS₂ (HhaI)	GCGC	>50	31
Haemophilus influenzae JC9	HinJCI (HindII)	GTPy↑PuAC	34	2
	HinJCII (HindIII)	AAGCTT	6	1
Haemophilus parahaemolyticus	HphI	GGTGA	>50	12
Haemophilus parainfluenzae	HpaI	GTT↑AȦC	13	0
	HpaII	C↑CGG	>50	26
Haemophilus suis	HsuI (HindIII)	A↑AGCTT	6	1
Halococcus agglomeratus	HagI	?	?	?
Herpetosiphon giganteus HP1023	HgiAI	G(A/T)GC(A/T)↑C	24	8
Herpetosiphon giganteus Hpg 5	HgiBI (AvaII)	G↑G(A/T)CC	>17	8
Herpetosiphon giganteus Hpg 9	HgiCI	G↑GPyPuCC	13	9
	HgiCII (AvaII)	G↑G(A/T)CC	>17	8
	HgiCIII (SalI)	G↑TCGAC	2	1
Herpetosiphon giganteus Hpa2	HgiDI (AcyI)	GPu↑CGPyC	>14	6
	HgiDII (SalI)	G↑TCGAC	2	1
Herpetosiphon giganteus Hpg 24	HgiEI (AvaII)	G↑G(A/T)CC	>17	8
	HgiEII	ACC(N)₆GGT	?	2
Herpetosiphon giganteus Hpg 14	HgiFI	?	?	?
Herpetosiphon giganteus Hpa 1	HgiGI (AcyI)	GPu↑CGPyC	>14	6
Herpetosiphon giganteus HP1049	HgiHI (HgiCI)	G↑GPyPuCC	>13	9
	HgiHII (AcyI)	GPu↑CGPyC	>14	6
	HgiHIII (AvaII)	G↑G(A/T)CC	>17	8
Herpetosiphon giganteus HFS101	HgiJI	?	?	?
	HgiJII	GPuGCPy↑C	7	2
Herpetosiphon giganteus Hpg 32	HgiKI	?	>18	?
Klebsiella pneumoniae OK8	KpnI	GGTAC↑C	2	0
Mastigocladus laminosus	MlaI (AsuII)	TT↑CGAA	7	0
Microbacterium thermosphactum	MthI (MboI)	GATC	>50	22
Micrococcus luteus	MluI	A↑CGCGT	7	0
Micrococcus radiodurans	MraI (SacII)	CCGCGG	4	0
Microcoleus species	MstI	TGC↑GCA	>10	4
	MstII (SauI)	CC↑TNAGG	2	0
Moraxella bovis	MboI	↑GATC	>50	22
	MboII	GAAGA	>50	11
Moraxella bovis	MbvI	?	?	?
Moraxella glueidi LG1	MglI	?	?	?
Moraxella glueidi LG2	MglII	?	?	?
Moraxella kingae	MkiI (HindIII)	AAGCTT	6	1
Moraxella nonliquefaciens	MnoI (HpaII)	C↑CGG	>50	26
	MnoII (MnnIII)	?	>10	?
	MnoIII (MboI)	GATC	>50	22
Moraxella nonliquefaciens	MnlI	CCTC	>50	26
Moraxella nonliquefaciens	MnnI (HindII)	GTPyPuAC	34	2
	MnnII (HaeIII)	GGCC	>50	22
	MnnIII	?	>10	?
	MnnIV (HhaI)	GCGC	>50	31
Moraxella nonliquefaciens	MniI (HaeIII)	GGCC	>50	22
	MniII (HpaII)	CCGG	>50	26
Moraxella osloensis	MosI (MboI)	GATC	>50	22

Microorganism	Enzyme	Sequence	Number of cleavage sites λ	Number of cleavage sites pBR322
Moraxella phenylpyruvica	MphI (EcoRII)	CC(^A_T)GG	> 35	6
Moraxella species	MspI (HpaII)	C↑CGG	> 50	26
Myxococcus stipitatus Mxs2	MsiI (XhoI)	CTCGAG	1	0
	MsiII	?	?	?
Myxococcus virescens V-2	MviI	?	1	?
	MviII	?	?	?
Neisseria caviae	NcaI (HinfI)	GANTC	> 50	10
Neisseria cinerea	NciI (CauII)	CC(^C_G)GGG	> 15	10
Neisseria denitrificans	NdeI	CA↑TATG	?	1
	NdeII (MboI)	GATC	> 50	22
Neisseria flavescens	NflI (MboI)	GATC	> 50	22
	NflII	?	?	?
	NflIII	?	?	?
Neisseria gonorrhoea	NgoI (HaeII)	PuGCGCPy	> 30	11
Neisseria gonorrhoea	NgoII (HaeIII)	GGCC	> 50	22
Neisseria gonorrhoea KH 7764-45	NgiIII (SacII)	CCGCGG	4	0
Neisseria mucosa	NmuI (NaeI)	GCCGGC	2	4
Neisseria ovis	NovI	?	?	?
	NovII (HinfI)	GANTC	> 50	10
Nocardia aerocolonigenes	NaeI	GCC↑GGC	2	4
Nocardia amarae	NamI (NarI)	GGCGCC	2	4
Nocardia argentinensis	NarI	GG↑CGCC	2	4
Nocardia blackwellii	NblI (PvuI)	CGAT↑CG	3	1
Nocardia brasiliensis	NbrI (NaeI)	GCCGGC	2	4
Nocardia brasiliensis	NbaI (NaeI)	GCCGGC	2	4
Nocardia corallina	NcoI	C↑CATGG	6	0
Nocardia dassonvillei	NdaI (NarI)	GG↑CGCC	2	4
Nocardia minima	NmiI (KpnI)	GGTACC	2	0
Nocardia opaca	NopI (SalI)	G↑TCGAC	2	1
	NopII	?	?	?
Nocardia otitidis-caviarum	NotI	?	0	0
Nocardia otitidis-caviarum	NocI (PstI)	CTGCAG	18	1
Nocardia rubra	NruI	TCG↑CGA	7	1
Nocardia uniformis	NunI	?	?	?
	NunII (NarI)	GG↑CGCC	2	4
Nostoc species	NspBI (AsuII)	TTCGAA	7	0
	NspBII	C(^A_T)G↑C(^T_C)G	?	21
Nostoc species	Nsp(7524)I	PuCATG↑Py	> 15	4
	Nsp(7524)II (SduI)	G(^G_A)GC(^C_A)↑C	?	10
	Nsp(7524)III (AvaI)	C↑PyCGPuG	8	1
	Nsp(7524)IV (AsuI)	G↑GNCC	> 30	15
	Nsp(7524)V (AsuII)	TTCGAA	7	0
Nostoc species	NspHI (NspCI)	PuCATG↑Py	> 15	4
	NspHII (AvaII)	GG(^A_T)CC	> 17	8
Oerskovia xanthineolytica	OxaI (AluI)	AGCT	> 50	16
	OxaII	?	?	?
Proteus vulgaris	PvuI	CGAT↑CG	3	1
	PvuII	CAG↑CTG	15	1
Providencia alcalifaciens	PalI (HaeIII)	GGCC	> 50	22
Providencia stuartii 164	PstI	CTGCA↑G	18	1
Pseudoanabaena species	PspI (AsuI)	GGNCC	> 30	15
Pseudomonas aeruginosa	PaeR7	?	1	0
Pseudomonas facilis	PfaI (MboI)	GATC	> 50	22
Pseudomonas maltophila	PmaI (PstI)	CTGCAG	18	1
Rhizobium leguminosarum 300	RleI	?	6	?
Rhizobium lupini 1	RluI	?	1	?

131

Microorganism	Enzyme	Sequence	Number of cleavage sites λ	pBR322
Rhizobium meliloti	*Rma*I	?	8	?
Rhodococcus rhodochrous	*Rrh*I (*Sal*I)	GTCGAC	2	1
	*Rrh*II	?	?	?
Rhodococcus rhodochrous	*Rro*I (*Sal*I)	GTCGAC	2	1
Rhodococcus species	*Rhs*I (*Bam*HI)	GGATCC	5	1
Rhodococcus species	*Rhp*I (*Sal*I)	GTCGAC	2	1
	*Rhp*II	?	?	?
Rhodococcus species	*Rhe*I (*Sal*I)	GTCGAC	2	1
Rhodospirillum rubrum	*Rrb*I	?	?	?
Rhodopseudomonas sphaeroides	*Rsp*I (*Pvu*I)	CGATCG	3	1
Rhodopseudomonas sphaeroides	*Rsh*I (*Pvu*I)	CGAT↑CG	3	1
Rhodopseudomonas sphaeroides	*Rsa*I	GT↑AC	> 50	3
Rhodopseudomonas sphaeroides	*Rsr*I (*Eco*RI)	GAATTC	5	1
Salmonella infantis	*Sin*I (*Ava*II)	GG(A̲/G)CC	>17	8
Serratia marcescens Sb	*Sma*I	CCC↑GGG	3	0
Serratia species SAI	*Ssp*I	?	?	?
Sphaerotilus natans C	*Sna*I	GTATAC	2	1
Spiroplasma citri ASP2	*Sci*NI (*Hha*I)	G↑CGC	> 50	31
Staphylococcus aureus 3A	*Sau*3A (*Mbo*I)	↑GATC	> 50	22
Staphylococcus aureus PS96	*Sau*96I (*Asu*I)	G↑GNCC	> 30	15
Staphylococcus saprophyticus	*Ssa*I	?	>10	?
Streptococcus cremoris F	*Scr*FI	CCNGG	> 50	16
Streptococcus durans	*Sdu*I	G(A̲/T)GC(A̲/T)C (G/C)	?	10
Streptococcus dysgalactiae	*Sdy*I (*Asu*I)	GGNCC	> 30	15
Streptococcus faecalis var. *zymogenes*	*Sfa*I (*Hae*III)	GG↑CC	> 50	22
Streptococcus faecalis GU	*Sfa*GUI (*Hpa*II)	CCGG	> 50	26
Streptococcus faecalis ND547	*Sfa*NI	GCATC	> 50	22
Streptomyces achromogenes	*Sac*I	GAGCT↑C	2	0
	*Sac*II	CCGC↑GG	4	0
	*Sac*III	?	> 100	?
Streptomyces albus	*Sal*PI (*Pst*I)	CTGCA↑G	18	1
Streptomyces albus subspecies *pathocidicus*	*Spa*I(*Xho*I)	CTCGAG	1	0
Streptomyces albus G	*Sal*I	G↑TCGAC	2	1
	*Sal*II	?	>20	?
Streptomyces aureofaciens IKA 18/4	*Sau*I	CC↑TNAGG	2	0
Streptomyces bobili	*Sbo*I (*Sac*II)	CCGCGG	4	0
Streptomyces caespitosus	*Sca*I	AGTACT	6	1
Streptomyces cupidosporus	*Scu*I (*Xho*I)	CTCGAG	1	0
Streptomyces exfoliatus	*Sex*I (*Xho*I)	CTCGAG	1	0
	*Sex*II	?	2	?
Streptomyces fradiae	*Sfr*I (*Sac*II)	CCGCGG	4	0
Streptomyces ganmycicus	*Sqa*I (*Xho*I)	CTCGAG	1	0
Streptomyces goshikiensis	*Sgo*I (*Xho*I)	CTCGAG	1	0
Streptomyces griseus	*Sgr*I	?	0	?
Streptomyces hygroscopicus	*Shy*TI	?	2	?
Streptomyces hygroscopicus	*Shy*I (*Sac*II)	CCGCGG	4	0
Streptomyces lavendulae	*Sla*I (*Xho*I)	C↑TCGAG	1	0
Streptomyces luteoreticuli	*Slu*I (*Xho*I)	CTCGAG	1	0
Streptomyces oderifer	*Sod*I	?	?	?
	*Sod*II	?	?	?
Streptomyces phaeochromogenes	*Sph*I	GCATG↑C	4	1
Streptomyces stanford	*Sst*I (*Sac*I)	GAGCT↑C	2	0
	*Sst*II (*Sac*II)	CCGC↑GG	4	0
	*Sst*III (*Sac*III)	?	> 100	?
	*Sst*IV (*Bcl*I)	TGATCA	7	0
Streptomyces tubercidicus	*Stu*I	AGG↑CCT	5	0

Microorganism	Enzyme	Sequence	Number of cleavage sites	
			λ	pBR322
Streptoverticillium flavopersicum	*Sfl*I (*Pst*I)	CTGCA↑G	18	1
Thermoplasma acidophilum	*Tha*I (*Fnu*DII)	CG↑CG	> 50	23
Thermopolyspora glauca	*Tgl*I (*Sac*II)	CCGCGG	4	0
Thermus aquaticus YTI	*Taq*I	T↑CGA	> 50	7
	*Taq*II	?	> 30	?
Thermus aquaticus	*Taq*XI (*Eco*RII)	CC↑AGG	> 35	6
Thermus flavus AT62	*Tfl*I (*Taq*I)	TCGA	> 50	7
Thermus thermophilus HB8	*Tth*HB8 I (*Taq*I)	TCGA	> 50	7
Thermus thermophilus strain 23	*Ttr*I (*Tth*111 I)	GACNNNGTC	2	1
Thermus thermophilus strain 110	*Tte*I (*Tth*111 I)	GACNNNGTC	2	1
Thermus thermophilus strain 111	*Tth*111 I	GACN↑NNGTC	2	1
	*Tth*111 II	CAAPuCA	> 30	5
	*Tth*111 III	?	?	?
Tolypothrix tenuis	*Ttn*I (*Hae*III)	GGCC	> 50	22
Vibrio harveyi	*Vha*I (*Hae*III)	GGCC	> 50	22
Xanthomonas amaranthicola	*Xam*I (*Sal*I)	GTCGAC	2	1
Xanthomonas badrii	*Xba*I	T↑CTAGA	1[d]	0
Xanthomonas holcicola	*Xho*I	C↑TCGAG	1	0
	*Xho*II	Pu↑GATCPy	> 20	8
Xanthomonas malvacearum	*Xma*I (*Sma*I)	C↑CCGGG	3	0
	*Xma*II (*Pst*I)	CTGCAG	18	1
	*Xma*III	C↑GGCCG	2	1
Xanthomonas manihotis 7AS1	*Xmn*I	GAANN↑NNTTC	> 11	2
Xanthomonas nigromaculans	*Xni*I (*Pvu*I)	CGATCG	3	1
Xanthomonas oryzae	*Xor*I (*Pst*I)	CTGCAG	18	1
	*Xor*II (*Pvu*I)	CGAT↑CG	3	1
Xanthomonas papavericola	*Xpa*I (*Xho*I)	C↑TCGAG	1	0

Restriction maps and DNA size markers

Plasmid pBr322

The total length of pBR322 is 4363 bp. The first T in the nucleotide sequence of the *Eco* RI site, GAATTC, is designated as nucleotide number 1. Restriction-enzyme site positions are given in numbers after the restriction enzyme position shown below and refer to the first nucleotide at the 5′ side of the recognition site. The position and direction of transcription of the ampicillin-resistance gene (Ap), the tetracycline-resistance gene (Tc) and the origin of replication (ORI) are shown (Fig. A2.1).

Sizes (in base pairs) of the fragments generated by digesting pBR322 with some common restriction enzymes. (From Sutcliffe (1978), *Nucleic Acids Research*, **5**, 2721)

Hae II	*Hae* III	*Taq* I	*Alu* I
1876	587	1444	910
622	540	1307	659
439	504	475	655
430	458	368	521
370	434	315	403
227	267	312	281
181	234	141	257
83	213		226
60	192		136
53	184		100
21	124		63
	123		57
	104		49
	89		19
	80		15
	64		11
	57		
	51		
	21		
	18		
	11		
	7		

Fig. A2.1. Restriction map of pBR322.

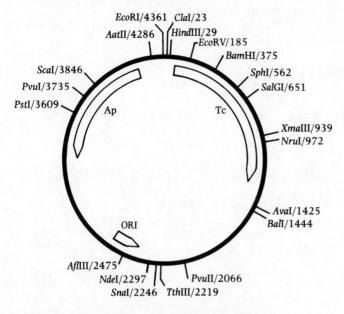

Fig. A2.2. Restriction map of M13mp8 and M13mp10.

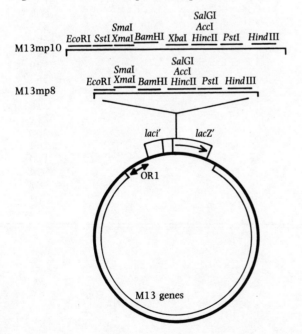

Bacteriophages M13mp8 and M13mp10

The total lengths of these two molecules are 7229 and 7245 bp. respectively. The multiple cloning sites within the *lacZ* gene are shown in expanded form (Fig. A2.2).

Bacteriophage λ Restriction fragment sizes (in base pairs) of λ *cI* *ind1-ts857 Sam7* (From Sanger *et al.* (1982), *Journal of Molecular Biology*, **162**, 729)

Eco RI	Hind III	Eco RI + Hind III	Bam HI	Bgl II	Pst I
21226	23130	21226	16841	22010	11501
7421	9416	5148	7233	13286	5077
5804	6682	4973	6770	9688	4749
5643	4361	4266	6527	2392	4507
4878	2322	3530	5626	651	2838
3530	2027	2027	5505	415	2556
	564	1584		60	2459
	125	1375			2443
		947			2140
		904			1986
		831			1093
		564			1700
		125			1159
					605
					514
					468
					448
					339
					264
					247
					216
					211
					200
					164
					150
					94
					87
					72
					15

Genetic nomenclature

The system of symbols used to represent the genotype of a particular
strain is often difficult for a novice, or even an experienced worker used
to the system for a different organism, to understand. Regrettably, there
is no standard nomenclature for all organisms and there remain cases of
inconsistency of usage within a given organism. We present here the
systems suggested for *E. coli* by Demerec *et al.* (1966) and for *S. cerevisiae*
by Sherman (1981) in the hope that they will become standard for all
prokaryotic and eukaryotic organisms.

E. coli

(Adapted from Demerec, Adelberg, Clark & Hartman, *Genetics*, **54**,
61–76, 1966.)

(i) Each genetic locus is designated by a three-letter, lower-case, italicised
symbol.

(ii) Different loci, any one of which may mutate to give the same gross
phenotype, are distinguished from each other by an italicised capital
letter added to the three-letter, lower-case symbol. For example, there are
a number of loci which encode enzymes involved in the biosynthetic
pathway for the amino acid, leucine. Mutations in any one of these loci
may give rise to strains which require the addition of leucine to the
medium in order to grow. Such mutants are said to have a leucine
auxotrophic phenotype. All the genetic loci involved are designated by
the three-letter symbol *leu*, the different individual loci being
distinguished as *leu A*, *leu B* etc.

(iii) When a mutation site is first defined it is given a serial identification
number after the locus symbol from which it is separated by a hyphen.
Thus a group of leucine auxotrophic mutants might be designated *leu-1*,
leu-2, *leu-3* etc. on first isolation. When the mutation can be assigned to a
particular genetic locus, then the hyphen is replaced by the locus letter,
e.g. *leuA1*, *leuC2*, *leuB3*. The isolation number, however, is not changed.

(iv) The wild-type allele of a given genetic locus is designated by the
superscript +, e.g. *leuB*$^+$. Mutant alleles are designated by their isolation
number alone, e.g. *leu-3* and not *leu-3*$^-$, *leuB3* and not *leuB3*$^-$. The
designation *leuB*$^-$ is not infrequently employed but its use is discouraged.

(v) The superscripts ^{R} and ^{S} for resistance and sensitivity are used to
define a strain's phenotype and not its genotype.

(vi) Plasmids and other extrachromosomal elements are designated by
symbols which are clearly distinguishable from those used for
chromosomal loci.

For naturally occurring plasmids, the first letter of the symbol is capitalised, the symbol is not italicised and the symbol is placed in brackets, e.g. (ColE1).

Recombinant plasmids are treated similarly but are designated by a lower-case p followed by two capital letters. The capital letters refer to the researcher who constructed the plasmid or the institution in which the work was performed, e.g. pSC101 (SC for Stanley Cohen) and pMT555 (MT for Manchester Technology). A register of these two-letter designations is now kept but the system is far from being standardised.

(vii) Insertion sequences are named IS1, IS2, IS3 etc. and transposons Tn1, Tn2, Tn3 etc. Some of the first transposons to be isolated were assigned a capital letter instead of a number. This letter indicates the particular drug resistance gene which the transposon carries, e.g. TnA carries an ampicillin-resistance gene.

The insertion of an IS element or transposon into the bacterial chromosome causes a mutation which can be mapped in the classical manner. The location and nature of such an insertion event is described by the following nomenclature. The standard symbol for the gene or region into which the element is inserted is followed by a hyphen and an isolation number in just the same way as any other mutant allele (see (iv) above). The isolation number is then followed by a double colon and the symbol for the inserted element. Thus an insertion of Tn501 into the *lacZ* gene might be represented as *lacZ*-15::Tn501.

(viii) Due to its importance in performing genetic crosses with *E. coli*, the F sex-factor has a nomenclature system all to itself.

F$^-$ A strain which lacks the F sex-factor.

F$^+$ A strain which harbours an autonomous F-factor which does not carry within it any portion of the *E. coli* chromosome which can be recognised genetically.

F′ (F-prime) A strain which harbours an autonomous F-factor which contains *E. coli* chromosomal sequences. This chromosomal segment confers upon the F′ plasmid some recognisable genotype.

HFr A strain which has an F-factor integrated into its chromosome. HFr stands for high-frequency recombination and indicates that the strain can efficiently donate chromosomal genes during conjugation. However many loci on the F-factor are involved in the determination of this property and therefore HFr, F$^+$, F$^-$ and F′ should be taken as simply indicating the molecular state of the F-factor and not any particular mating activity.

(ix) The phenotype of a strain is often designated by a three-letter symbol which is not italicised and has a capitalised first letter, e.g. Tet^R and Tet^S for resistance and sensitivity to the antibiotic tetracycline; Leu$^+$ and Leu$^-$ for leucine prototrophy and auxotrophy. These phenotypic designations should be explained when first used in any publication. The following examples provide a guide to the use of genetic symbols in *E. coli*:

leuB   a genetic locus
leuB$^+$ a wild-type allele of that locus
leuB3   a mutant allele
leu-3   a leucine auxotrophic mutation as yet unassigned to a particular locus
Leu$^+$ a strain which does not require leucine in its growth medium (a leucine prototroph)
Leu$^-$ a strain which cannot grow unless leucine is included in its growth medium (a leucine auxotroph).

S. cerevisiae

(Adapted from Sherman, F. In *The Molecular Biology of the Yeast* Saccharomyces: *life cycle and inheritance*, ed. J. N. Strathern, E. W. Jones & J. R. Broach, pp. 639–40. New York: Cold Spring Harbor Laboratory, 1981.)

The system for *S. cerevisiae* is broadly based on the proposals of Demerec *et al.* (1966) for bacteria. Certain changes have been made to avoid the use of superscripts, notably the use of upper-case symbols for wild-type alleles. The system for the designation of extrachromosomal loci is more refined and a dual system is in use to name recombinant plasmids.

(i) Genes are designated by a three-letter, italicised symbol which should, whenever possible, be consistent with the proposals for bacteria. Thus all genetic loci which may mutate to give leucine auxotrophs are designated *LEU*. Particular loci are distinguished by a number immediately following the three-letter symbol, e.g. *LEU1*, *LEU2*, *LEU3* etc. This locus number is used universally and is given by the workers who first defined a particular locus. Alleles are designated by a second number, separated from the locus number by a hyphen, e.g. *leu2-112*. These allele numbers are peculiar to individual laboratories.

(ii) Gene clusters, complementation groups within a gene or separable domains within a gene are distinguished by a capital letter following the locus number, e.g. *his4A*, *his4B*, *his4C*.

(iii) Dominant and recessive alleles of genes are denoted by upper-case and lower-case letters respectively, e.g. *LEU2* and *leu2*. Frequently the dominant allele is also the wild-type form of the gene, as is the case for the *LEU2* locus. However, sometimes it is the mutant allele which is dominant. This is the case for suppressor loci, e.g. *SUP6* (mutant) and *sup6* (wild-type) and dominant resistances, e.g. *CUP1* (copper-resistant mutant, dominant) and *cup1* (copper-sensitive wild-type, recessive).

(iv) Wild-type genes may be designated simply as + when it is obvious which gene is referred to. This convention is most often used in representing the genotype of diploids, e.g.

MATa	ade1	ade2	+	gal2
MATα	+	+	leu2	+

The + symbol may also be used to identify specific genes as wild-type. This may be used to emphasise that it is the recessive allele which is the wild-type, e.g. *sup6* +, *cup1* +.

(v) The system avoids the use of superscripts but genes conferring resistance or sensitivity can have their alleles distinguished by a superscript ^{R} or ^{S} respectively, e.g. *can^{R}1*, *CAN^{S}1*, *CUP^{R}1*, *cup^{S}1*.

(vi) The method of indicating a strain's phenotype is the same as that used for bacteria. Thus Leu$^+$ and Leu$^-$ designate leucine prototrophy and auxotrophy respectively.

(vii) Extrachromosomal (non-Mendelian) genes use the same symbol system as chromosomal genes but they are enclosed in square brackets, e.g. [*oli1 var1*]. The presence or absence of the endogenous 2μ plasmid is indicated as [circ$^+$] or [circ^o]. The system avoids the use of Greek symbols and where they have been employed historically they are transliterated into Roman script, e.g. *psi$^+$* and *psi$^-$* for ψ$^+$ and ψ$^-$.

(viii) There are two parallel systems for naming recombinant plasmids. The first uses the pXY123 system employed in bacterial genetics. The second attempts to indicate the type of molecule by a capital Y (for yeast) followed by either C, E, I or R to indicate yeast centromeric, episomal, integrative or replicative plasmids respectively. The second capital letter is followed by a lower-case p (for plasmid) and a number indicating the particular plasmid, e.g. YIp5, YRp7 etc. A full definition of YCp, YEp, YIp and YRp is given in the body of this Dictionary.

Genetic maps

The genetic linkage map of *Escherichia coli* (Fig. A4.1) is taken from Bachmann (1983), *Microbiology Reviews*, **47**, 180–230.

The *Bacillus subtilis* map (Fig. A4.2) is from Wilson (1982). In *Genetic Maps*, ed. S. J. O'Brien, pp. 133–49. Frederick, Maryland: National Cancer Institute.

The *Saccharomyces cerevisiae* map (Fig. A4.3) is from Mortimer & Schild (1982). In *The Molecular Biology of the Yeast* Saccharomyces: *Metabolism and Gene Expression*, ed. J. N. Strathern, E. W. Jones & J. R. Broach, pp. 639–50. New York: Cold Spring Harbor Laboratory.

Fig. A4.1. Genetic map of *Escherichia coli*.

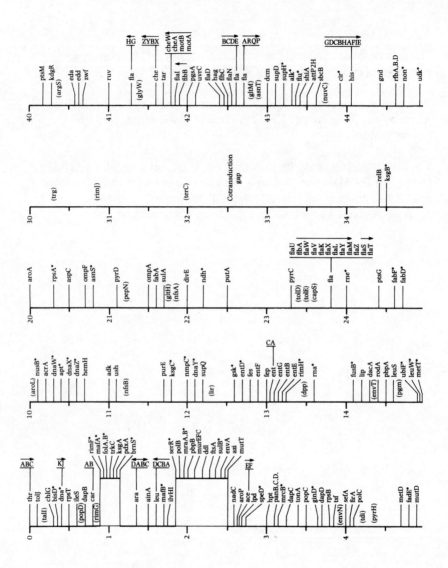

142

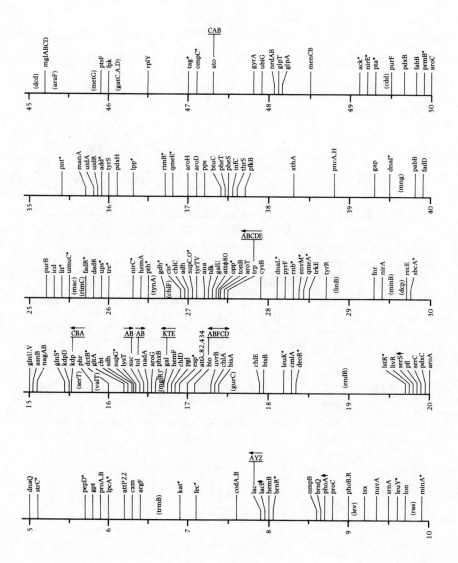

Genetic map of *Escherichia coli* (*cont.*).

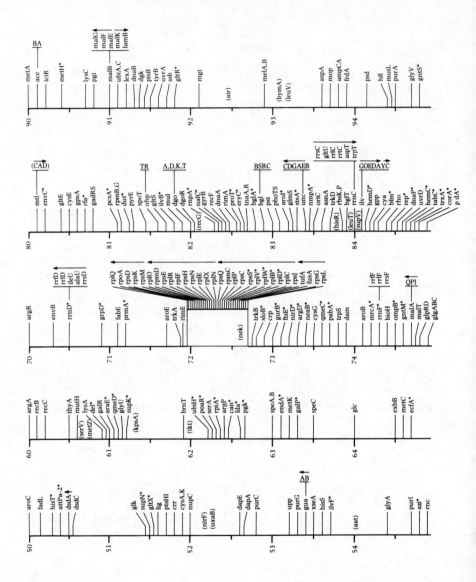

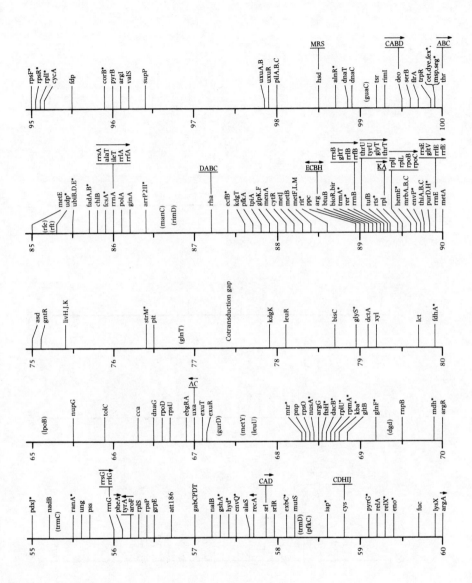

95

rpsF*
rpsR*
rplI*
cycA

fdp

corB*
pyrB
argI
valS

supP

uxuA,B
uxuR
pltA,B,C

hsd

MRS

alnR*
dnaT
dnaC

(guaC)

tsr
rimI

deo
serB
flrA
trpR

CABD

cet.dye.fex*
msp.seg
thr

ABC

96   97   98   99   100

85

(rfe)
(rft)

metE
udp*
ubiB,D,E*

fadA,B*
chlB
fcsA*
rrnA
polA
ginA

rrsA
alaT
ileT
rrfA
rrfA

arrP2II*

(manC)
(rimD)

rha

DABC

ecfB*
kdgT
pfkA
tpiA
glpK,F
menA
cytR
metJ
metB
metF,L,M
rit*
ppc
arg
btuB
bioR.bir

ECBH

trmA*
rer*
rrnB

rrsB
gltT
rrlB
rrfB

tufB
rts*
rpl

thrU
tyrU
glyT
thrT

rplJ

KA

hemE*
mrhA,B,C
envD*
thiA,B,C
purD,H*
rrnE
metA

rpl
rplL
rpoB
rpoC*

rrsE
gltV
rrlE
rrfE

86   87   88   89   90

75

asd
gntR

livH,J,K

strM*
pit

(glnT)

kdgK

leuR

Cotransduction gap

bisC

glyS*
dctA
xyl

lct

ldhA*

76   77   78   79   80

65

(lpoB)

nupG

tolC

cca

dnaG
rpoD
rpsU

ebgRA
uxa
exuT
exuR

ΔC

(gurD)

(metY)

(leuU)

mtr*
pnp
rpsO
nusA*
argG
ftsH*
dacB*
rplU*
kba*
gltB

glnF*

(dgd)

rmpB

mdh*
argR

66   67   68   69   70

55

pdxJ*
nadB

(trmC)

ranA*
ung
pss

rrnG

rrsG
rrlG

pheA
tyrA
aroF
rplS
rpsP
grpE

att186

gabCPDT

nalB
gshA*
hyd*
envQ*
alaS
recA

CAD

srl
strR

exbC*
mutS

(trmD)
(pfkC)

CDHIJ

iap*

cys

pyrG*
relA
relX*
eno*

fuc

lysX
argA

56   57   58   59   60

145

Fig. A4.2. Genetic map of *Bacillus subtilis*.

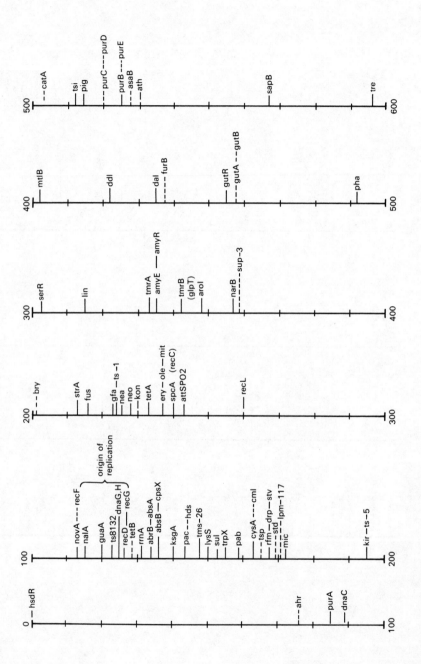

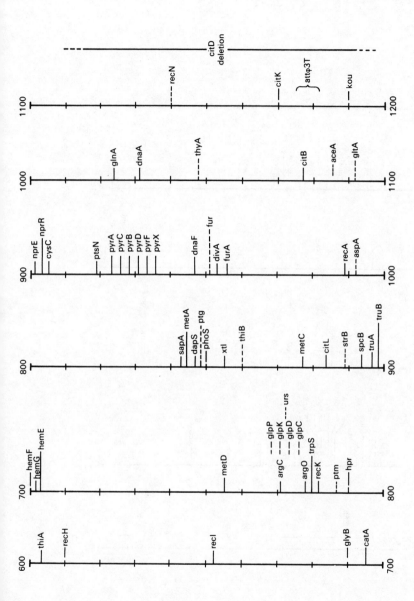

147

Genetic map of *Bacillus subtilis* (*cont.*).

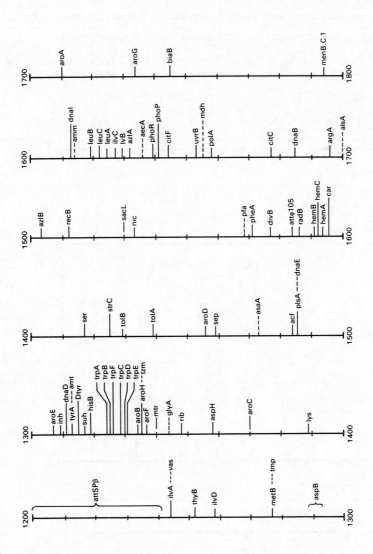

148

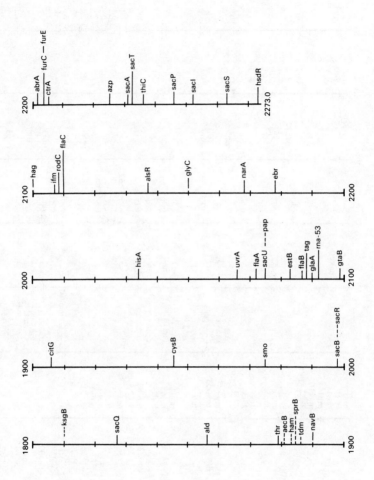

Fig. A4.3. Genetic map of *Saccharomyces cerevisiae*.

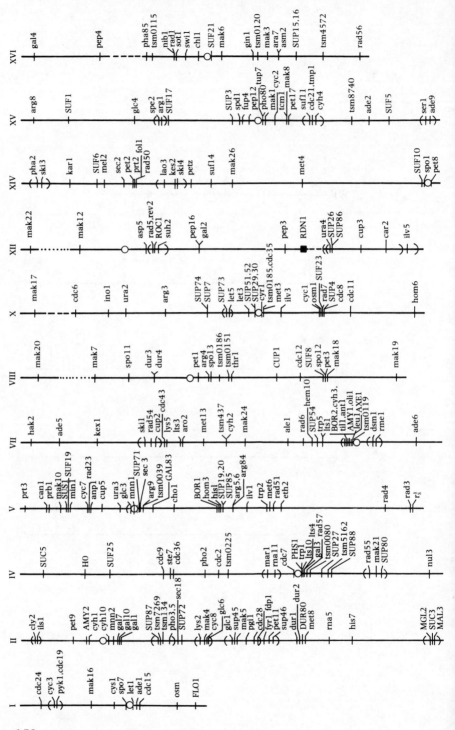

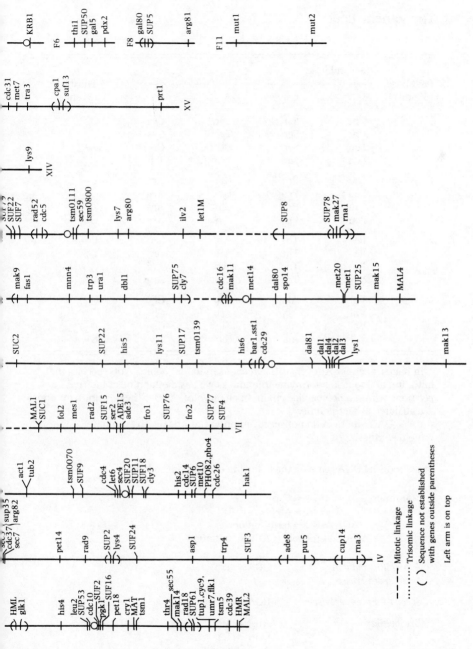

XBL829-4132A

- - - - Mitotic linkage
· · · · · · Trisomic linkage
( ) Sequence not established
    with genes outside parentheses

Left arm is on top

151

The genetic code

First base (5' end)	Second base				Third base (3' end)
	U	C	A	G	
U	Phe	Ser	Tyr	Cys	U
	Phe	Ser	Tyr	Cys	C
	Leu	Ser	OCHRE†	OPAL†	A
	Leu	Ser	AMBER†	Trp	G
C	Leu	Pro	His	Arg	U
	Leu	Pro	His	Arg	C
	Leu	Pro	Gln	Arg	A
	Leu	Pro	Gln	Arg	G
A	Ile	Thr	Asn	Ser	U
	Ile	Thr	Asn	Ser	C
	Ile	Thr	Lys	Arg	A
	Met*	Thr	Lys	Arg	G
G	Val	Ala	Asp	Gly	U
	Val	Ala	Asp	Gly	C
	Val	Ala	Glu	Gly	A
	Val*	Ala	Glu	Gly	G

* In bacteria, either AUG or GUG can act as initiator codons. AUG specifies the initiating amino acid n-formylmethionine as well as methionine at internal positions within a polypeptide chain. In eukaryotes, AUG is the only initiator and is translated as methionine.

† UAA, UAG and UGA do not specify amino acids and act to terminate protein synthesis.

Mitochondrial DNAs employ the following exceptional codons:

Mammals
- UGA = Trp
- AUA = Met
- AGA, AGG = termination
- AUA and (possibly) AUU = initiation

Aspergillus nidulans
Neurospora crassa
- UGA = Trp

Saccharomyces cerevisiae
- UGA = Trp

Drosophila
- AUA = Met
- UGA = Trp
- AGA (possibly) = Ser

Zea mays
- UGA = Trp
- CGC = Trp

The one- and three-letter abbreviations for amino acids

Amino acid	Three-letter abbreviation	One-letter abbreviation
Alanine	Ala	A
Arginine	Arg	R
Asparagine	Asn	N
Aspartic acid	Asp	D
Cysteine	Cys	C
Glutamic acid	Glu	E
Glutamine	Gln	Q
Glycine	Gly	G
Histidine	His	H
Isoleucine	Ile	I
Leucine	Leu	L
Lysine	Lys	K
Methionine	Met	M
Phenylalanine	Phe	F
Proline	Pro	P
Serine	Ser	S
Threonine	Thr	T
Tryptophan	Trp	W
Tyrosine	Tyr	Y
Valine	Val	V
Termination	—	*